8th International Conference on Mechatronics and Control Engineering (ICMCE 2019)

IOP Conference Series: Materials Science and Engineering
Volume 707

Paris, France
23 - 25 July 2019

ISBN: 978-1-7138-1037-7
ISSN: 1742-6588

Printed from e-media with permission by:

Curran Associates, Inc.
57 Morehouse Lane
Red Hook, NY 12571

Some format issues inherent in the e-media version may also appear in this print version.

Institute of Physics
Dirac House, Temple Back
Bristol BS1 6BE UK

Phone: 44 1 17 929 7481
Fax: 44 1 17 920 0979

techtracking@iop.org

Additional copies of this publication are available from:

Curran Associates, Inc.
57 Morehouse Lane
Red Hook, NY 12571 USA
Phone: 845-758-0400
Fax: 845-758-2633
Email: curran@proceedings.com
Web: www.proceedings.com

TABLE OF CONTENTS

CONTROL ENGINEERING

MATERIAL SCIENCE

IOP Conf. Series: Materials Science and Engineering **707** (2019) 011001 doi:10.1088/1757-899X/707/1/011001

Preface

It is our pleasure to welcome you to the 2019 8th International Conference on Mechatronics and Control Engineering (ICMCE 2019) in Paris, France on July 23-25, 2019. ICMCE 2019 is a conference dedicated to topics in acoustics and noise control, computational mechatronics and engineering design management, mechatronics, process control & instrumentation, automated guided vehicles and solid mechatronics, nanomaterial engineering etc.

The proceedings present a selection of 19 papers submitted to the conference from universities, research institutes and industries. All of the papers were subjected to peer-review by conference committee members and international reviewers. Papers included in the proceedings have been selected according to quality and relevance to the conference themes. The proceedings aim to present to the readers recent advances in the field of mechatronics and control engineering and in various related areas.

We would like to express our genuine gratitude to everyone who has contributed to ICMCE 2019. The opening remark including a warm welcome to all participants has been given by Prof. Jan Awrejcewicz, The Lodz University of Technology, Poland. Our especially thanks to the conference chair, Prof. Giuseppe Conte, Università Politecnica delle Marche, Italy, the program chairs, Prof. Olivier Sename, Grenoble Institute of Technology, France, Assoc. Prof. Ke-Lin Du, Concordia University, Canada and Prof. Changguo Wang, Harbin Institute of Technology, China, all the technical committee members. Let us wish ICMCE the same success for next year.

Conference Chairs
Prof. Jan Awrejcewicz, The Lodz University of Technology, Poland
Prof. Giuseppe Conte, Università Politecnica delle Marche, Italy

IOP Conf. Series: Materials Science and Engineering **707** (2019) 011002 doi:10.1088/1757-899X/707/1/011002

Conference Committees

Conference Chairs

Jan Awrejcewicz, The Lodz University of Technology, Poland

Magnus Thor Jonsson, University of Iceland, Iceland

Giuseppe Conte, Università Politecnica delle Marche, Italy

Program Chairs

Ke-Lin Du, Concordia University, Canada

Olivier Sename, Grenoble Institute of Technology, France

Changguo Wang, Harbin Institute of Technology, China

Technical Committees

Mehmet Emir KOKSAl, Ondokuz Mayis University, Turkey

Xiaobing Mao, Wuhan University of Technology, China

Parvaneh Esmaili, Near East University, Cyprus

Zareena Kausar, Air University, Pakistan

N. Mohana Sundara Raju, Mahendra Institute of Technology, India

Karina R. G. Silva, Escola de Engenharia Elétrica, Mecânica e de Computação, Brazil

Jacob Scharcanski, Universidade Federal do Rio Grande do Sul, Brazil

Mohamed Ichchou, Ecole Centrale de Lyon, France

Xiaolin Wang, Engineering and Technology University of Tasmania,Australia

Dario Mendoza Chipantasi, Universidad de las Fuerzas Armadas ESPE, Ecuador

Gokhan Erdemir, Istanbul Sabahattin Zaim University, Turkey

Emin Taner ELMAS, Iskenderun Technical University, Turkey

Mingbiao Yu, Southeast University, China

Sarunya Promkotra, Khon Kaen University, Thailand

Sharifuddin Mondal, National Institute of Technology Patna, India

Peer review statement

All papers published in this volume of *IOP Conference Series: Materials Science and Engineering* have been peer reviewed through processes administered by the proceedings Editors. Reviews were conducted by expert referees to the professional and scientific standards expected of a proceedings journal published by IOP Publishing.

IOP Conf. Series: Materials Science and Engineering **707** (2019) 012001 doi:10.1088/1757-899X/707/1/012001

Adaptive unknown input observer for nonlinear systems with application to fault diagnosis

Sharifuddin Mondal[1]

[1]Department of Mechanical Engineering, National Institute of Technology Patna, Bihar, PIN-800005, India (e-mail: sharifuddin@nitp.ac.in).

Abstract. An adaptive unknown input observer for nonlinear systems whose nonlinear functions satisfy Lipschitz condition is presented. The main advantage of the proposed observer is that it is sufficient to satisfy the standard existence conditions of normal unknown input observer design. Another advantage is that it estimates both states and parameters simultaneously. So it can be very much useful for robust control and fault diagnosis. As a possible application, an actuator fault diagnosis scheme is developed for single link robot arm. Finally, numerical simulation is performed to validate the effectiveness of the proposed observer in estimating the parameter along with states.

1. Introduction

The importance of unknown input observer (UIO) is already well established for their diverse applications in robust control and fault diagnosis. Over the years, many types of unknown input observers have been proposed for both linear and nonlinear systems [1-6]. With the increasing demands for high precision control techniques, importance of unknown input observer that is robust and/or adaptive is increasing. In recent years, some researches have been carried out to design UIOs that are adaptive against certain parameters [7-12]. Most of these works are related to designing adaptive unknown input observer (AUIO) either for linear systems or application specific nonlinear systems. Mondal [12] developed an AUIO for state estimations of certain type of nonlinear systems. But there are no such AUIO that works for general types of nonlinear systems.

In this work, an adaptive unknown input observer for nonlinear systems has been developed. The observer is designed on the assumption that nonlinear functions of the system satisfies well know Lipschitz condition. The sufficient conditions of the observer design are also formulated. The main advantages of the proposed observer are that it is enough to satisfy the conditions that are required to design a conventional unknown input observer. Then the AUIO is reformulated considering a special scenario that is required to apply the observer to single link robot arm. This kind of observers finds wide applications in the fields of adaptive control, robust control and fault diagnosis [13-15]. As a possible application, an actuator fault diagnosis method for single link robot arm is outlined. Then numerical simulation is performed using MATLAB/SIMULINK software to demonstrate the ability of the observer in estimating states and parameters simultaneously. In this work, a fault is introduced in a parameter. By comparing estimated parameter with its nominal values, fault is detected and isolated.

2. Adaptive unknown input observer

2.1. Problem statement

Consider a nonlinear system with unknown inputs as

$$\dot{x}(t) = Ax(t) + f(x(t), u(t)) + Bg(x(t), u(t))\theta + Ed(t) \tag{1}$$

$$y(t) = Cx(t) \tag{2}$$

where $x(t) \in R^n$, $u(t) \in R^m$, $y(t) \in R^p$ $d(t) \in R^q$ are state, input, output and unknown input vectors. The matrices A, B, E, C are known matrices with appropriate dimensions. The functions $f(x(t), u(t))$ and $g(x(t), u(t))$ are continuous functions. Before defining the observer, following assumptions are stated.

Assumption 1: The unknown parameter θ is bounded for a constant σ as

$$|\theta|_2 < \sigma. \tag{3}$$

Assumption 2: The nonlinear functions $f(x(t), u(t))$ and $g(x(t), u(t))$ are globally Lipschitz, i.e.,

$$\left\|f\big(x(t), u(t)\big) - f\big(\hat{x}(t), u(t)\big)\right\| \leq \gamma_1(x(t) - \hat{x}(t)) \tag{4}$$

$$\left\|g\big(x(t), u(t)\big) - g\big(\hat{x}(t), u(t)\big)\right\| \leq \gamma_2(x(t) - \hat{x}(t)) \tag{5}$$

where γ_1 and γ_2 are Lipschitz constants.

Assumption 3: For a symmetric positive definite matrix P_s, a matrix S can be found out by satisfying

$$(PB)^T P_s = SC. \tag{6}$$

The assumptions 1-2 are standard assumptions considered in designing observers. Assumption 3 is also widely used in designing adaptive observers [10, 12].

Now an unknown input observer for the above system is designed as

$$\dot{z}(t) = Nz(t) + Ly(t) + Pf(\hat{x}(t), u(t)) + PBg(\hat{x}(t), u(t))\hat{\theta} \tag{7}$$

$$\hat{x}(t) = z(t) - Hy(t) \tag{8}$$

$$\dot{\hat{\theta}}(t) = \frac{g\big(\hat{x}(t), u(t)\big)^T S e_y(t)}{\rho} \text{ with } \rho > 0. \tag{9}$$

where $e_y(t) = y(t) - \hat{y}(t)$ is output error, $\hat{x}(t) \in R^n$, $\hat{y}(t) \in R^p$ are estimated state and output vectors respectively. The matrices N, L, P, H, S are unknown matrices that need to be determined.

Consider the state error as

$$\begin{aligned} e(t) &= x(t) - \hat{x}(t) \\ &= x(t) - z(t) + HCx(t). \text{ (using eq. (8))} \\ &= Px(t) - z(t) \end{aligned} \tag{10}$$

where

$$P = I + HC. \tag{11}$$

The error dynamics $\dot{e}(t)$ of the above observer (7)-(9) for the system (1) and (2) can be written as

$$\dot{e}(t) = PAx(t) + Pf\big(x(t), u(t)\big) + PBg\big(x(t), u(t)\big)\theta + PEd(t) - N\big(\hat{x}(t) + HCx(t)\big) - LCx(t) - Pf\big(\hat{x}(t), u(t)\big) - PBg\big(\hat{x}(t), u(t)\big)\hat{\theta}$$

Assuming $\tilde{\theta} = \theta - \hat{\theta}$ and simplifying the above equation, one can write as

$$\dot{e}(t) = Ne(t) + (PA - LC - NP)x(t) + P\Big(f\big(x(t), u(t)\big) - f\big(\hat{x}(t), u(t)\big)\Big) + PEd(t) + PB\Big(g\big(x(t), u(t)\big) - g\big(\hat{x}(t), u(t)\big)\Big)\theta + PBg(\hat{x}(t), u(t))\tilde{\theta} \tag{12}$$

Now following conditions are applied to remove the unmeasurable variables:

$$PA - LC - NP = 0 \tag{13}$$

$$PE = 0. \tag{14}$$

Now assume

$$K = L + NH. \tag{15}$$

From equations (11) and (13), one gets

$$N = PA - KC \tag{16}$$

Using equations (13), (14) and (16), one can write the equation (12) as

$$\dot{e}(t) = (PA - KC)e(t) + P\Big(f\big(x(t), u(t)\big) - f\big(\hat{x}(t), u(t)\big)\Big) + PB\Big(g\big(x(t), u(t)\big) - g\big(\hat{x}(t), u(t)\big)\Big)\theta + PBg(\hat{x}(t), u(t))\tilde{\theta} \tag{17}$$

Now the problem in hand is that the observer gain matrices should be determined in such a way that the observer error dynamics (17) become stable i.e., $e(t) \to 0$ as $t \to \infty$. The method of determining the observer matrices is presented in the next subsection.

2.2. Main results

In this subsection, the method of determining the observer gain matrices (N, L, P, H, S) is discussed. The gain matrix H is calculated using equations (11) and (14) as

$$H = -E(CE)^{+} + Y(I_p - (CE)(CE)^{+}) \tag{18}$$

Once H is known, P is determined using equation (11). To determine the other matrices(N, L), following theorem is proposed.

Theorem 1: For a given system, represented by equations (1)-(2), the system, defined by equations (7)-(9), will be the adaptive unknown input observer if there exists a symmetric matrix P_s and the LMI

$$\begin{bmatrix} P_s(PA) - ZC + (PA)^T P_s - C^T Z + 2I & P_s & P_s B \\ P_s & -\varepsilon_1 I & 0 \\ B^T P_s & 0 & -\varepsilon_2 I \end{bmatrix} < 0$$

where $\varepsilon_1 = (\gamma_1^2 \sigma_m^2)^{-1}$ and $\varepsilon_2 = (\gamma_2^2 \sigma_m^2 \sigma^2)^{-1}$ holds. The gain matrix $K = (P_s)^{-1} Z$ also stabilizes the error dynamics (17), i.e., $e(t) \to 0$ as $t \to \infty$.

To prove this theorem, following lemma is required.

Lemma 1: If the nonlinear functions $f(x(t), u(t))$ and $g(x(t), u(t))$ satisfy conditions (4) and (5) respectively, then following inequalities hold.

$$2e^T(t)P_sP\left(f(x(t),u(t)) - f(\hat{x}(t),u(t))\right) \le \gamma_1^2\sigma_m^2 e^T(t)P_sP_se(t) + e^T(t)e(t) \tag{19}$$

$$2e^T(t)P_sPB\left(g(x(t),u(t)) - g(\hat{x}(t),u(t))\right)\theta \le \gamma_2^2\sigma_m^2\sigma^2 e^T(t)P_sBB^TP_se(t) + e^T(t)e(t) \tag{20}$$

where σ_m is norm value of the matrix P.

Proof: This can be proved in the same way that is presented in [10] for similar cases. Proof is ignored here due to space limitation.

Proof of the theorem 1: Consider Lyapunov-Krasovskii functional $V(t)$ as

$$V(t) = e^T(t)P_se(t) + \rho\tilde{\theta}^T\tilde{\theta} \tag{21}$$

where $\rho > 0$ and P_s is symmetric positive definite matrix

Differentiating (21) with respect to time, one gets

$$\mathrm{V}(t) = 2e^T(t)P_s\dot{e}(t) + 2\rho\tilde{\theta}^T\dot{\tilde{\theta}} \tag{22}$$

Replacing (17) in (22), one can write

$$\dot{V}(t) = e^T(t)\{P_s(PA - KC) + (PA - KC)^TP_s\}e(t) + 2e^T(t)P_sP\left(f(x(t),u(t)) - f(\hat{x}(t),u(t))\right) + 2e^T(t)P_sPB\big(g(x(t),u(t))\theta - g(\hat{x}(t),u(t))\hat{\theta}\big) + 2\rho\tilde{\theta}^T\dot{\tilde{\theta}} \tag{23}$$

Now after rearrangement and applying lemma 1, one can write

$$\dot{V}(t) \le e^T(t)\{P_s(PA - KC) + (PA - KC)^TP_s\}e(t) + \gamma_1^2\sigma_m^2 e^T(t)P_sP_se(t) + e^T(t)e(t) + \gamma_2^2\sigma_m^2\sigma^2 e^T(t)P_sBB^TP_se(t) + e^T(t)e(t) + 2e^T(t)P_sPBg(\hat{x}(t),u(t))\tilde{\theta} + 2\rho\tilde{\theta}^T\dot{\tilde{\theta}} \tag{24}$$

To discard the effect of unknown parameter from equation (24), following condition is applied.

$$2e^T(t)P_sPBg(\hat{x}(t),u(t))\tilde{\theta} + 2\rho\tilde{\theta}^T\dot{\tilde{\theta}} = 0. \tag{25}$$

This relation gives the form required for parameter estimation given in (9) as θ is generally constant.

Now the relation (24) is written as

$$\dot{V}(t) = e^T(t)\{P_s(PA - KC) + (PA - KC)^TP_s + \gamma_1^2\sigma_m^2P_sP_s + \gamma_2^2\sigma_m^2\sigma^2P_sBB^TP_s + 2I\}e(t) \tag{26}$$

It is well established that if $\dot{V}(t) < 0$, then $e(t) \to 0$ as $t \to \infty$. Now from equation (26), $\dot{V}(t) < 0$ is possible only when following condition is satisfied.

$$P_s(PA - KC) + (PA - KC)^TP_s + \gamma_1^2\sigma_m^2P_sP_s + \gamma_2^2\sigma_m^2\sigma^2P_sBB^TP_s + 2I < 0 \tag{27}$$

Finally applying Schur's complement [2, 10, 16] in (27), one gets the form presented in theorem 1. This completes the proof.

Now solving the LMI presented in theorem 1 using LMI toolbox provided with MATLAB software package [17], P_s and Z are determined. Using these values, K is then calculated as $K = (P_s)^{-1}Z$. The matrix N is determined using the relation (16). The matrix L is found out from (15). Finally the matrix S is chosen arbitrarily satisfying the condition (6). This completes the observer design. In the next section, the above result will be reformulated for a special type of nonlinear system.

3. AUIO when g(.) depends only on inputs

In this section, the adaptive observer is reformulated for special type of nonlinear system. The system equation is written as follows.

$$\dot{x}(t) = Ax(t) + f(x(t), u(t)) + Bg(u(t))\theta + Ed(t) \tag{28}$$

Now the problem is to design an adaptive UIO for the system represented by equations (28) and (2). Here the nonlinear function $g(.)$ depends only on input $u(t)$. The assumptions (1) and(3) will remain the same but little change will occur in assumption (2) as only one nonlinear function will be there instead of two. Now the observer for the system (28) and (2) can be written as

$$\dot{z}(t) = Nz(t) + Ly(t) + Pf\big(x(t), u(t)\big) + PBg(u(t))\hat{\theta} \tag{29}$$

$$\hat{x}(t) = z(t) - Hy(t) \tag{8}$$

$$\dot{\hat{\theta}}(t) = \frac{g(u(t))^T S e_y(t)}{\rho} \text{ with } \rho > 0. \tag{30}$$

where $e_y(t) = y(t) - \hat{y}(t)$ and equation (8) is rewritten for better understanding.

Assuming the state error like previous case, the error dynamics for the presented system can be determined as

$$\dot{e}(t) = (PA - KC)e(t) + P(f\big(x(t), u(t)\big) - f\big(\hat{x}(t), u(t)\big) + PBg(u(t))\tilde{\theta} \tag{31}$$

In the same way as previous section, following theorem is presented to determine the observer matrices.

Theorem 2: For a given system, represented by equations (28) and (2), the system, defined by equations (29), (8) and (30), will be the adaptive unknown input observer if there exists a symmetric matrix P_s and the LMI

$$\begin{bmatrix} P_s(PA) - ZC + (PA)^T P_s - C^T Z + I & P_s \\ P_s & -\varepsilon_1 I \end{bmatrix} < 0$$

where $\varepsilon_1 = (\gamma_1^2 \sigma_m^2)^{-1}$ holds. The gain matrix $K = (P_s)^{-1} Z$ also stabilizes the observer error dynamics (31), i.e., $e(t) \to 0$ as $t \to \infty$.

Proof of theorem 2: Considering the Lyapunov-Krasvoskii functional as given in (21) and following the steps mentioned in the proof of theorem 1, one can easily prove the theorem 2. So detailed proof is omitted here.

Other observer matrices are determined in the same way that are presented in the previous section. A numerical example is presented in the next section to show the effectiveness of the observer.

4. Numerical example-AUIO in fault diagnosis

In this section, a single link robot arm that is widely used for validating different estimation and control algorithms [2-3, 10, 12] is considered. First, the actuator fault diagnosis method is presented. Then simulation results are given.

4.1 Actuator Fault Diagnosis

In state space model, any actuator fault can be accommodated into input matrix (B). Let the matrix B becomes B_u when a fault occurs. In the present problem (single link robot arm), single actuator is present, so a constant δ that relates these two matrices can be considered as

$$B_u = B\delta \tag{32}$$

In ideal situation, $\delta = 1$ (no fault). When fault occurs, it will be changed to another value depending on level of fault. The idea of fault diagnosis using adaptive observer is that this δ is assumed as adaptive parameter i.e., $\theta = \delta$ and the system is remodelled. Then an adaptive observer is designed to estimate the states and parameters. Other uncertain parameters are incorporated as unknown inputs so their effects will not appear in estimated states and parameters. When a fault occurs, the estimated parameter will settle to a value other than 1. So observing the estimated parameter, fault can be detected. Magnitude of the fault can also be determined using the nominal values of input matrix and using relation (32).

4.2 Numerical results

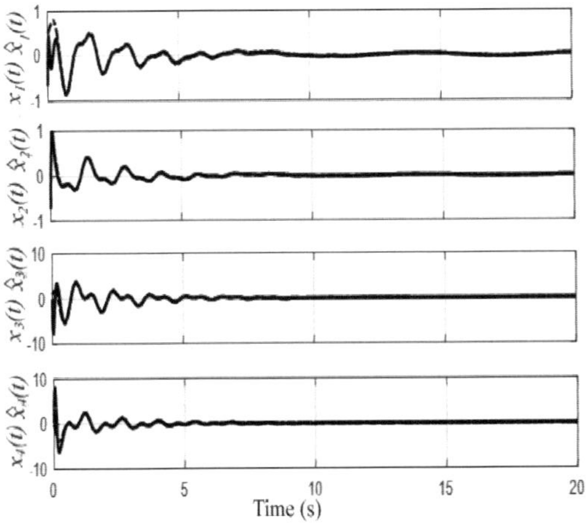

Fig. 1: True and estimated states (solid-true states and dash/dot-estimated states)

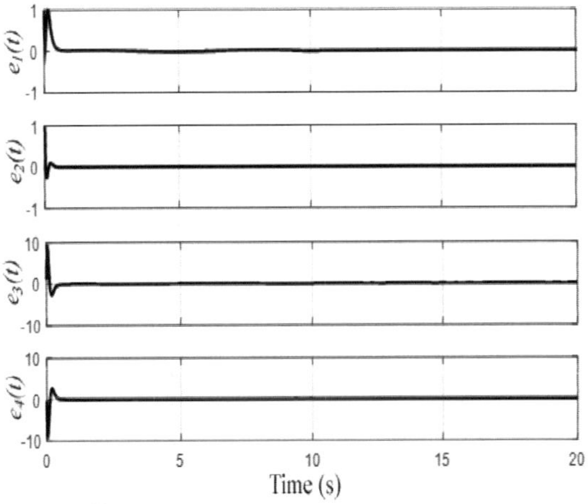

Fig. 2: State errors

It is assumed that the value of the viscous friction constant F_2 is uncertain and unknown. The effect of this parameter in the system is considered to have a known part ($F_2\dot{\theta}_2(\mathrm{t})$) and an unknown part which is modelled as an unknown input $d(\mathrm{t})$. It is assumed that a fault occurs in the actuator. So the value of B becomes unknown (B_u). It is remodelled as $B_u = B\theta$. In ideal situation, $\|\theta\| = 1$, which gives $\sigma = 1$. With this arrangement, the system matrices and vectors are given as follows:

$$A = \begin{bmatrix} 0 & 0 & 1 & 0 \\ 0 & 0 & 0 & 1 \\ -48.6486 & 48.6486 & -1.2432 & 0 \\ 19.3548 & -19.3548 & 0 & -0.6882 \end{bmatrix}, \quad B = \begin{bmatrix} 0 \\ 0 \\ 2.1622 \\ 0 \end{bmatrix}, \quad E = \begin{bmatrix} 0 \\ 0 \\ 0 \\ 1 \end{bmatrix}, \quad d(t) = -\frac{\Delta F_2 \dot{\theta}_2(t)}{J_2},$$

$$g\big(x(t), u(t)\big) = \begin{bmatrix} 0 \\ 0 \\ 0 \\ -33.2274\cos(x_2) \end{bmatrix} \text{and } x(t) = \begin{Bmatrix} x_1 \\ x_2 \\ \dot{x}_1 \\ \dot{x}_2 \end{Bmatrix}.$$

The measurement matrix is taken as $C = \begin{bmatrix} 0 & 1 & 0 & 0 \\ 0 & 0 & 1 & 1 \end{bmatrix}$, which satisfies the existence conditions of adaptive unknown input observer. The observer matrices P, H are determined with a suitable Y. Then the LMI technique is used to determine other matrices.

The final observer matrices are $P = \begin{bmatrix} 1 & 0.8147 & 0 & 0 \\ 0 & 1.9058 & 0 & 0 \\ 0 & 0.1270 & 1 & 0 \\ 0 & 0.9134 & -1 & 0 \end{bmatrix}$, $H = \begin{bmatrix} 0.8147 & 0 \\ 0.9058 & 0 \\ 0.1270 & 0 \\ 0.9134 & -1 \end{bmatrix}$,

$$N = \begin{bmatrix} 0 & -33.7675 & -13.4126 & -13.5979 \\ 0 & -38.6494 & -9.3419 & -7.4361 \\ -48.6486 & 342.1144 & 103.0181 & 104.3883 \\ 48.6486 & -351.0777 & -123.0050 & -123.3348 \end{bmatrix}, \quad L = \begin{bmatrix} 89.6360 & 0.8147 \\ 81.6360 & 1.9058 \\ -672.1426 & 0.1270 \\ 709.0683 & 0.9134 \end{bmatrix}, \quad S = \begin{bmatrix} 4.4295 & 1.2090 \end{bmatrix}, \ \rho = 0.04.$$

The input signal for the system is chosen as follows: $\boldsymbol{u} = 1.0\sin(wt)$ with $w = 1$ rad / s . The fault is introduced in such a way that θ changes to 0.5 at 10 seconds from 1. Now the adaptive observer is designed. The true & estimated states and the estimated errors are plotted in Fig 1 and Fig 2 respectively.

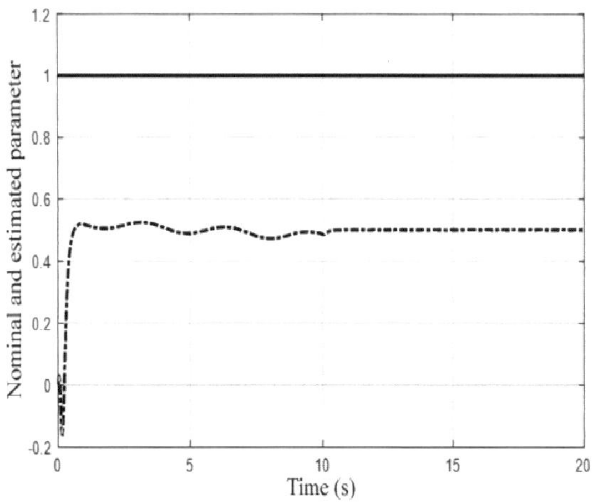

Fig. 3: Nominal and estimated parameter (solid-nominal and dash/dot-estimated)

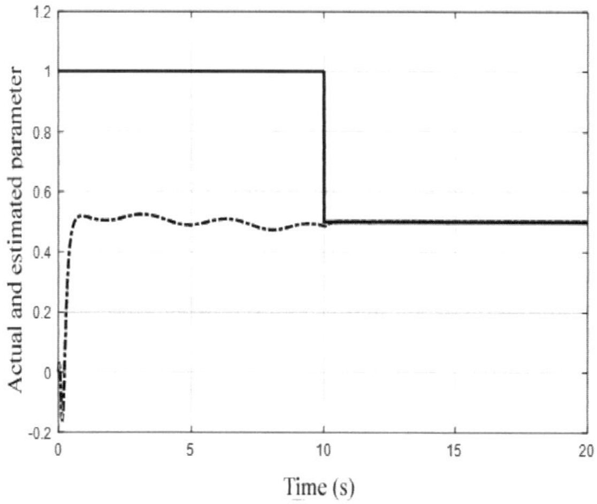

Fig. 4: Actual and estimated parameter (solid-actual and dash/dot-estimated)

Nominal and estimated parameters are plotted in Fig 3. Actual and estimated parameters are shown in Fig. 4. Fig. 1 and Fig. 2 show that the observer is estimating the states efficiently. Fig. 3 confirms that the estimated parameter is settling at 0.5 other than 1 that is confirming the fault in the actuator. Fig 4 confirms the efficacy of the observer in estimating the parameters correctly. As the observer is designed for constant parameters, it is not able to tract the changes from 1 to 0.5.

5. Conclusion

An adaptive unknown input observer has been presented. The observer is designed for nonlinear systems that satisfy Lipschitz condition. This type of observer is very much useful in adaptive and robust controls. It can estimate both states and parameters simultaneously. So it is very useful for fault diagnosis of nonlinear systems. As a possible application, an actuator fault diagnosis method has been devised for single link robot arm. Using numerical simulation, effectiveness of the proposed observer is validated.

6. References

[1] Ahmadizadeh S, Zarei J and Karimi HR 2014 *Asian J. Contr.* **16 (4)** 1009-1019.
[2] Mondal S 2015 *Int. J. Dyn. Contr.* **3(4)** 448-456.
[3] Mondal S, Chakraborty G and Bhattacharyya K 2007 *Proc. Int. Conf. Adv. on Contr. Optim. Dyn. Sys.* (Bangalore) 117-120.
[4] Ali AA 2013 *Ph.D. dissertation* (University of Michigan, USA).
[5] Zhu F 2007 *IEEE Int. Conf. on Contr.Autom.* (Guangzhou) 529–534.
[6] Na J, Chen AS, Herrmann G, Burke R and Brace C 2018 *IEEE Trans. on Veh. Tech.* **67(1)** 409-421.
[7] Bowong S and Tewa JJ 2008 *Math. Comp. Modell.* **48(11-12)** 1826-1839.
[8] Heydari M and Demetriou MA 2017 *Inter. J. Syst. Sci.* **48(1)** 182-189.
[9] Lin S-Y, Yen J-Y, Chen M-S, Chang S-H and Kao C-Y 2017 *IEEE Trans. on Autom. Contr.* **62(8)** 4073-4078.
[10] Mondal S and Chung WK 2013 *Inter. J. Adapt. Contr. Sig. Proc.* 27(7) 610-619.
[11] Salgado I and Chirez I 2018 *IEEE Trans. on Neu. Netw. Learn. Syst.* 29(8) 3499-3509.
[12] Mondal S 2018 *Inter. Conf. on Recent Adv. in Mat. Manuf. Tech.* (Hyderabad).
[13] Guo R, Guo K, Gen Q, Zhang J, Dong J and Bai L 2016 *Math. Prob. Engg.* **2016** 1-12.
[14] Lee H, Snyder S and Hovakimyan N 2014 *AIAA Guid. Nav. Contr. Conf.* (National Harbor, Maryland).
[15] Wang D and Lum K-Y 2007 *Int. J. Adapt. Contr. Sig. Proc.* **21(1)** 31–48.

IOP Conf. Series: Materials Science and Engineering **707** (2019) 012001 doi:10.1088/1757-899X/707/1/012001

[16] Boyd S, El Ghaoui L, Feron E and Balakrishnan V 1994 *Linear Matrix Inequalities in Systems and Control Theory* (SIAM: Philadelphia, PA).
[17] Gahinet P, Nemirovsky A, Laum AJ and Chilali M 1995 *LMI Control Toolbox: For Use with MATLAB* (The Mathworks Inc.: USA).

IOP Conf. Series: Materials Science and Engineering **707** (2019) 012002 doi:10.1088/1757-899X/707/1/012002

Design of control system for cover washing machine in the automotive manufacturing industry

Syahril Ardi[1] and Fariz Risky Pratama[2]

[1,2]Production Engineering & Manufacturing Processes, Polytechnic Manufacture Astra
Jl. Gaya Motor Raya No.8, Sunter II, Jakarta 14330, Indonesia

Abstract. In this paper, it discusses research conducted in the automotive manufacturing industry. In this case that is in the process of manufacturing the two-wheeled vehicle parts cover. In manufacturing the cover parts includes several processes, namely: Melting, Casting, Finishing, Machining, Painting, and Assembling. The washing process is the process of cleaning the chips and coolants that attach to the part, namely by spraying water and wind. The previous washing process used a manual method, namely by dipping the part in a container filled with water and then spraying it clean free of coolant. This method still has drawbacks, namely that there are still coolant marks left in the part. Coolant that has been attached to the part will be difficult to clean when it has dried. We conduct research by designing a cover washing machine that is integrated with the PLC (Programmable Logic Controller). In this case, the control system has 60 Input-Output, consisting of 36 inputs and 24 outputs. This machine construction uses the Rotary Table principle, so that the operator's work becomes more effective. In addition, the results reduce the amount of rejection in the painting process by 5% and increase the amount of production from 400 pieces per day to 700 pieces per day.

1. Introduction

This research was conducted in the automotive manufacturing industry engaged in manufacturing aluminum automotive components. The growth of the automotive industry is currently very rapid, where cars and motorbikes have become a public need for transportation facilities. To meet the needs of society, quality and reliable automotive components are needed. In addition to having to comply with safety standards, it must also provide comfort when used.

One of the products produced is Cover, which are the crank case protective component and the engine on two-wheeled vehicles. Cover is made through several processes including: Melting, Casting, Finishing, Machining, Painting, and Assembling. In the Machining process there is one process, namely washing or commonly called Part washing before being taken to the next process.

In the previous, washing process used the manual method by dipping the part in a container filled with water and then spraying it clean free of coolant. But this method still has many disadvantages, namely that there are still coolant marks left in the part. Coolant will leave spots on the part if it dries. Considering after the washing process is the process of painting or painting if there is still coolant left behind, automatically the painting results become uneven. This research also aims to reduce the amount of rejection or defects caused by machines or Man Power. How is to design a cover washing machine control system by automatically [1-3].

2. Methodology and design

This company is a company engaged in the manufacture of automotive products ranging from 2W (2 wheels) and 4W (4 wheels) with the Die Casting method. One of the products produced is Cover.

Figure 1 shows an illustration of the process of manufacturing Cover products. The Cover Washing Machine is in the Machining process.

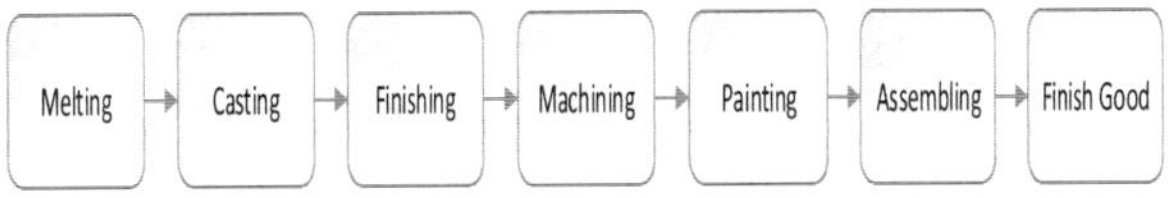

Figure 1. The cover manufacturing process

Based on industry demands, there are two conditions in the cover manufacturing process, namely: the first is the need for additional production, one of which is by adding new machines, turning machines from manuals into automatic, and by reducing cycle time. Secondly, it is necessary to change the engine. In this case, the existing machine cannot support the production target [4-5].

2.1. Machine Work Process

In summary, the cover washing machine work process includes:

- Manual Loading, which is the process of placing parts on washing machine jigs. Water Blow, Air Blow, Flush on-off, which is: the water starts to be sprayed, followed by air spraying, after a few seconds Flush will light up for a few seconds and then die.
- Rotary Table On-Off, that is: the jig table will start rotating again still counter clockwise and will stop after the sensor detects.
- Gripper On- Off, namely: in the third process here the gripper will drop back to hold the part, then the Air Blade will light up to make sure the part is completely clean, then the gripper will go back up and the motor starts to spin again.
- Unloading, namely: after the part returns to its original place, it's time for the operator to take the washed part and replace the next part.

2.2. Design of the machine

Figure 2 shows the parts in the Cover Washing machine, as follows:

1. Three Phase Motor, it used to rotate the rotary table.
2. Push Button, it is to run the machine manually.
3. Rotary Table is a place to put parts to be washed so as not to shake.
4. The main panel is a panel that contains controls as well as other electrical devices.
5. The control panel, it used to facilitate movement, can also change the mode on the machine.

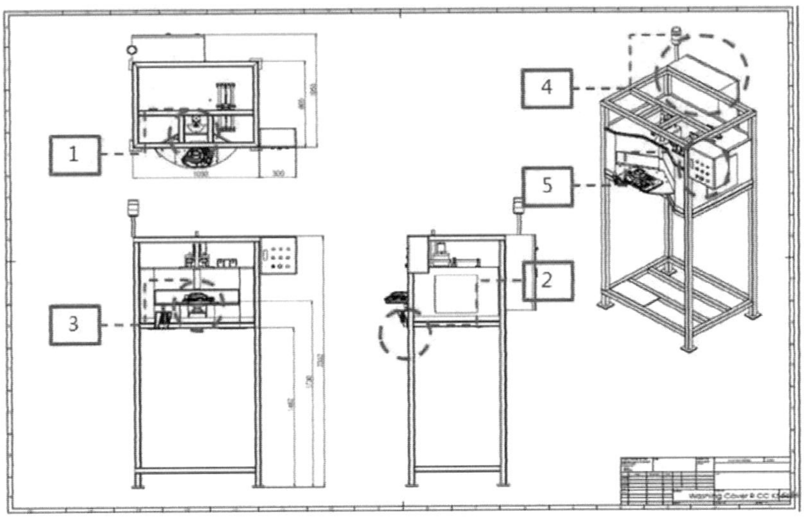

Figure 2. The parts in the washing cover machine

IOP Conf. Series: Materials Science and Engineering **707** (2019) 012002 doi:10.1088/1757-899X/707/1/012002

2.3. Control System Design

Figure 3 shows a block diagram of the system control design. This block diagram is an overview of the communication process on the washing cover machine control system. PLC receives the input signals from input devices in the form of switches, buttons, and proximity sensors. PLC provides a signal to the relay to activate the valve solenoids and the cylinders. PLC communicates to give signals to the inverter to run an electric motor [6].

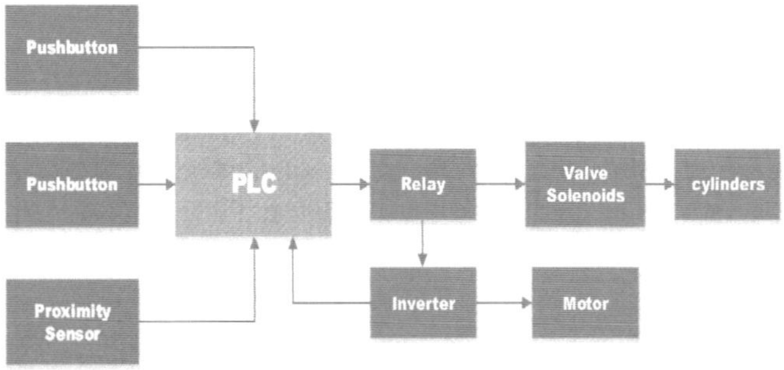

Figure 3. The diagram block of control system

3. Research and Discussion

In the process of manufacturing the control system this time is divided into 2 parts, namely electric wiring and design programming. Layout or placement of the components to be used must pay attention to the strategic and economic side; therefore the box panel is used as the placement of these components. This washing machine is equipped with two box panels, which consist of the main panel and the control panel. Figure 4 shows the realization of the cover washing machine and its description.

Figure 4. Cover Washing Machine

Description of cover washing machine:
1. Rotary Table. The rotating table is driven by a motor, functioning as a place to put parts.
2. Proximity Sensor. Proximity sensor detects rotation on the rotary table.
3. Lifter Nozzle. Rotary table retention time stops rotating.
4. Safety. If there is a hand that passes through the sensor, the machine will automatically stop.
5. Push Button & Emergency Stop. To start or stop the engine

In this Cover Washing Machine there are 3 main rooms, namely:
- Room 1, it is used for Loading and Unloading parts, can be done manually by the operator or done by a robot.
- Room 2, it is used to clean parts by spraying water and air.
- Room 3, it is used to make sure parts are clean or not, using a tool called a water blade.

Figure 5 shows the comparison of manual machines with automatic cover washing machines.

Figure 5. The comparison of manual machines with automatic cover washing machines.

3.1. Design of Programming

Programming on the Cover Washing machine uses hardware and software. The laptop is used as a hardware device in which the CX-Programmer software is installed. Figure 5 shows the flow chart of the cover washing machine work process. The program is divided into several work processes in one cycle and is automatic.

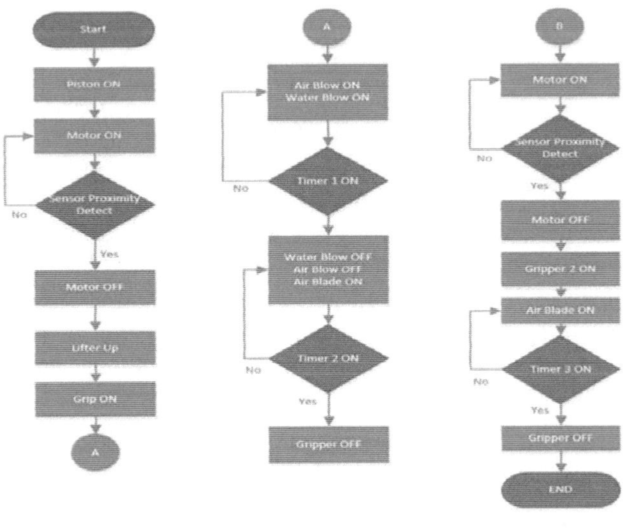

Figure 6. Flow Chart Program

The program is divided into several work processes in one cycle and is automatic. To run the auto program you must first press bstart1 and bstart2 buttons which have input addresses 1.02 and 1.04 which function to rotate the rotary table. After the auto process starts, the rotary table will rotate. After the sensor at the input address 2.07 works, the rotary table will stop and the stopper will go up and activate op1. At the input of 2.07 a div up is used which if the sensor detects many times but the data is sent only once.

The Rotary Ready program is used to ensure that the rotary table is ready to run or stop. This program is done to make it easier to detect the work of the rotary table, if there is one process that has not been running, the rotary table will not be able to turn on. The Rotary Table Start program is used to run W10.01 relay. If the relay is active, the rotary table will run. The Rotary Program Table run is a program used to run a motor on the rotary table. After being marked by W10.01 that is on, the motor will run. The function of the 011 timer is to make a pause or delay for 1 second.

After the Rotary Table rotates and the limit switch detected, the water spray will light up, provided the op1 is running. Water Spray is controlled by a counter and each sequence will move as many numbers as indicated by the counter. After that the water spray will stop and continue with the Air Spray. Similar to the Water Spray, the Air Spray is controlled using a counter and will move as much as printed on the counter. The difference is the delay using a timer which serves to provide a lag time so that the spraying of the wind carried out is more controlled and provides cleaner results. Manual programs are used to test each machine part before starting the machine automatically. The manual program uses two start & stop buttons, and a selector switch located on the front panel.

3.2. Results Analysis

By designing a cover washing machine, it has an impact on the reject reduction in the painting section by 5%. Part that is not clean from coolant if left for too long will cause mold that is difficult to clean and will result in difficult to process into painting. Also the parts will avoid the chips resulting from cutting machining, if not cleaned, the parts will be dirty because many chips are left on the part and have the same effect, which is difficult to paint. The Cover washing machine here also succeeded in adding production because the cycle time decreased from 50 second to 40 second and it produced 700 pieces per day from the previous one which could only produce 400 pieces per day. The dimensions of the machine are also in accordance with the layout available at The Company which has a length of 1.3 meters, a width of 1.05 meters, and a height of 2.32 meters. A high upward shape can save available space and is easier to operate by the operator. The standard part used also meets the requirements of The Company by using parts that are easy to find and use brands that have been used frequently such as Omron, Fuji Electric, Schneider so that machine maintenance becomes easier.

4. Conclusions

In this paper we have discussed the research on Control System Design of Cover Washing Machine With PLC-based in the Automotive Manufacturing Industry. The wiring on the machine panel is carried out by industry standards, namely in the form of marking on each cable used. The machine program uses the Rotary Table principle or rotating jig table. The resulting work effectiveness of the operator increased, thereby reducing the number of reject paintings by 5% and increasing the production of 400 pieces per day to 700 pieces per day. Cover Washing Machine control system is made with dimensions of 1,350 mm x 1,055 mm x 2,332 mm using Omron CP1E PLC as controller, proximity sensor as input and 3 Phase Motor as output. The control system has 60 Input-Outputs, consisting of 36 inputs and 24 outputs. In next research, a seal can be added to the door, so that the water inside is not spilled out. A review of the duration of spraying is needed to be completely clean from the coolant.

5. References

[1] Ardi, S., D Ardyansyah, *Design Control Systems of Human Machine Interface in the NTVS-2894 Seat Grinder Machine to Increase the Productivity*, IOP Conference Series: Materials Science and Engineering 306 (1), 012112, 2018.

[2] Ardi, S., A Ponco, RA Latief, *Design of integrated SCADA systems in piston production manufacturing case study on the conveyor, the coolant, the hydraulic, and the alarm systems using PLC*, IEEE Xplore, 187 – 191, 2017.

[3] Ardi, S., H Abdurrahman, *Design of pokayoke systems to increase the efficiency of function check oxygen sensor machine using programmable logic controller in manufacturing industry*, IEEE Xplore, 192 – 196, 2017.

[4] Alphonsus, E.R., Abdullah, M.O., A review on the applications of programmable logic controllers (PLCs), *Renewable and Sustainable Energy Reviews* 60 (2016), 1185-1205,.

[5] Gökhan Gelena, Murat Uzamb, The synthesis and PLC implementation of hybrid modular supervisors for real time control of an experimental manufacturing system, *Journal of Manufacturing Systems* 33 (2014) 535–550.

[6] Ardi, S., Defi, W.Y., *Control systems modification of loading and unloading in oil filling machine based on programmable logic controller for manufacturing industry*, AIP Conf. Proc. 2021, 060029-1–060029-6 (2018); https://doi.org/10.1063/1.5062793.

Grey box identification and adaptive control in a water level system

R Rodriguez[1], V Paredes[1], G Noa[2] and L Trucios[1]

[1] Department of Mechanical Engineering, National University of Engineering, Peru
[2] Department of Engineering, Pontifical Catholic University of Peru, Peru

Abstract. We study the application of Adaptive Control to estimate external disturbances in a Water Level System model, taking advantage of a constructed nonlinear grey box identification that allows to exploit its dynamics. The external disturbances will be specially related to valve manipulations considering limited sensing information using only water height sensors. Our methodology is twofold: (1) Using a grey box model to identify the plant under an operating point and, (2) using adaptive control to react against changes in the system or perturbations on the valve and still lead to a convergent and asymptotically stable controller.

1. Introduction

Water Level Systems are a common equipment in many industries, although PLC with built-in classical controllers are enough for many applications [1], sometimes system disturbances are mild and therefore intelligent controllers need to be used. When dealing with uncertainties it is common to use a variable structure controller or adaptive control scheme, while controllers like Sliding Mode Control are robust, they generate large control efforts in comparison to adaptive controllers. An adaptive algorithm can be used to tune gains of the controller [2] or to construct a learning law to compensate for the unknown parameters in the plant. Many approaches use ARMA or other black box models to identify a plant; however, using those approaches make a designer diverge from his or her physical interpretation and intuition of the model. If a parameter changes, it is not trivial to map it to the generated model. We propose to use a Grey Box Identification algorithm in order to preserve the integrity of the model and its relationship with the parameters. Once the identification of this plant with a defined structure is realized, varying parameters can be recognized due to unmodeled dynamics or physical changes in the system introduced externally. For instance, in a water level system, the hydraulic resistance is characterized by the slope of the pressure change with respect to the mass flow rate, which renders a nonlinear relationship that changes depending on its operating point [3]. To estimate the parameters to aid in the control action, we propose to use a Lyapunov based adaptive controller [4]. Adaptive schemes have been proposed for water level systems, although several of them do not exploit the problem structure [5,6] and sometimes do not use a Lyapunov verification of stability. We provide a method based on Lyapunov stability, which exploits the structure of the water level dynamics to guarantee asymptotic convergence of the system response to the reference point.

2. Water Level System

A plant corresponding to a two-tank system is shown in figure 1(a), this water level system is well known [3] and can be modelled as two tanks connected through water pipes as shown in figure 1(b). Tank 1 serves as a water supplier for tank 2 and it is refilled through a pump before starting operation.

IOP Conf. Series: Materials Science and Engineering **707** (2019) 012003 doi:10.1088/1757-899X/707/1/012003

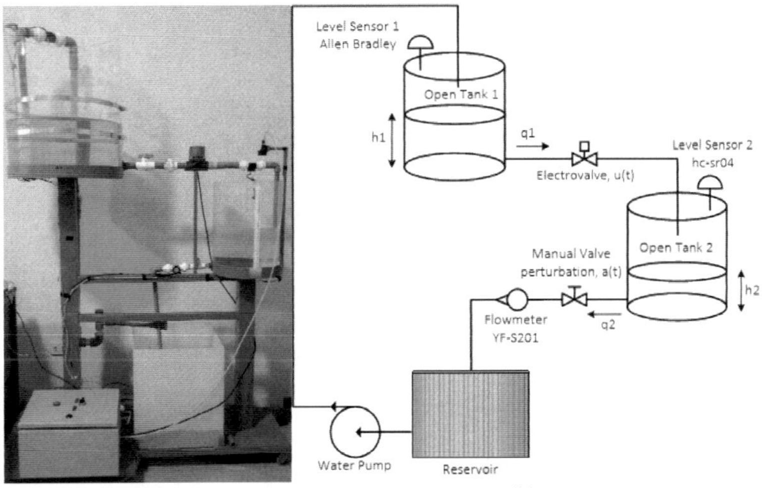

(a) (b)

Figure 1. (a) Physical Water Level System implementation. (b) Model of the two-tank water level system

The tanks have certain water heights (h_1, h_2) in the cylindrical reservoirs with areas A_1 and A_2. An electric pump generates the initial conditions of the tank 1. A solenoid valve regulated by a voltage $V(t)$ controls the input flow $q_1(t)$ of the tank 2; the output flow $q_2(t)$ of tank 2 can be affected by the manual valve 2 that is considered as a system perturbation. The measured signals are the water level h_1 and h_2 of both tanks and the water flow $q_2(t)$.

2.1. System Modelling

Using Bernoulli's law, a relationship between any two reference sections in the water can be expressed according to:

$$gz_1 + \frac{p_1}{\rho} + \frac{v_1^2}{2} = gz_2 + \frac{p_2}{\rho} + \frac{v_2^2}{2} \quad (1)$$

Where, v_k is the fluid speed in the considered section k, ρ is the fluid density, p_k is the pressure along the flow in the section k, g is the gravity and z_k is the height of the water column with respect to the reference for $k \in \{1,2\}$

The subscripts indicates the top and bottom levels respectively. The relationships between the flow rates q_1, q_2 and the water levels h_1, h_2 are as follows.

$$q_1 = c_1\sqrt{2gh_1} = k_1\sqrt{h_1(t)}u(t) \quad (2)$$

$$q_2 = c_2\sqrt{2gh_2} = k_2\sqrt{h_2(t)}a \quad (3)$$

Where c is the discharge coefficient of the valve [2], we assume a model where the input voltage directly affects the flow passing from tank 1 to tank 2. Also, k_1 and k_2 are gains that are defined to obtain a desired restriction ([0 - 1]) for a(t) and u(t), the perturbation and the input signal respectively. Hence, the relation between the flow rates q_1, q_2 and the dynamic system are as follows.

$$\frac{d}{dt}h_1(t) = -\frac{q_1}{A_1} \quad (4)$$

$$\frac{d}{dt}h_2(t) = \frac{q_1}{A_2} - \frac{q_2}{A_2} \quad (5)$$

By combining (2), (3) and (4), (5), the mathematical model of the system is expressed as follows.

$$\frac{d}{dt}h_1(t) = -\frac{k_1\sqrt{h_1(t)}u(t)}{A_1} \tag{6}$$

$$\frac{d}{dt}h_2(t) = \frac{k_1\sqrt{h_1(t)}u(t)}{A_2} - \frac{k_2\sqrt{h_2(t)}a}{A_2} \tag{7}$$

Where u(t) is a mapped function for the input voltage V(t) which operates in the range $[v_0 - V_{max}]$. The relation between u(t) and v(t) is as follows.

$$u(t) = \frac{V(t) - v_0}{V_{max} - v_0} \tag{8}$$

In this work, the constant parameters (k_1 , k_2 , v_0) are estimated using the Grey System Identification, through a non-linear model (6), (7). Also a(t) is a mapped function for the aperture of the valve (figure 1(b)), that is unknown and must be estimated by the algorithm in order to drive the system to the required set-point.

2.2. System setup

The water level system has a central processing unit based on Arduino Due with a 32-bit Atmel SAM3X8E ARM Cortex-M3 microcontroller. In this work, pulse width modulation (PWM) signal is converted to DC voltage that is used as an input of the solenoid valve. This valve provides control action on the system and operates in the range 5.7 - 10 VDC.

As shown in figure 1(b), two sensors are used in the system. A sensor Allen Bradley (871TM-B5N18-H2) measures the water level in tank 2 and an ultrasonic sensor (HC-SR04) is used to measures the water level for tank 1. Additionally, we use a water flow sensor (YF-S201) located in series in the output pipe of tank 2. This sensor is used to verify the estimation process.

3. Nonlinear grey box identification

Linear systems models can describe real models quite well, even when many systems are in fact, non-linear. The reason is that some non-linear systems can be approximated by linear models when the inputs and outputs are within certain ranges. However, sometimes-linear models are unable to capture the system behaviour, especially when operating in a broad range of operating points where non-linear phenomena occur.

If we understand the physics of a system and we can express it as a series, of Ordinary Differential Equations (ODE) we can represent the model with states and known and unknown parameters in the continuous time domain (or discrete time difference equations). With this representation we can use a grey box nonlinear identification. Grey box modelling is useful when you know the relation between variables, constraints or explicit equations that represent the dynamics of the system.

In our case, we use a grey box model in continuous time, for data acquired experimentally using the MATLAB software.

3.1. Nonlinear physical models

Non-linear state space models in continuous time are described by:

$$\begin{aligned} \dot{x}(t) &= f(x(t), u(t), \theta) \\ y(t) &= h(x(t), u(t), \theta) + e(t) \end{aligned} \tag{9}$$

Where $e(t)$ is white noise, $x(t)$ the states of the system, $u(t)$ is the input signal, $y(t)$ is the system response and θ the unknown parameters to be fitted. In a similar manner we define the estimated values ($\hat{x}$, $\hat{y}$, $\hat{\theta}$):

$$\dot{\hat{x}}(t) = f(\hat{x}(t), u(t), \theta)$$

$$\hat{y}(t|\theta) = h(\hat{x}(t), u(t), \theta) + e(t) \tag{10}$$

The objective of the identification is minimizing the prediction error using the cost function.

$$J(\theta) = \frac{1}{N}\sum_{i=1}^{N}\frac{1}{2}\left(y(t_i) - \hat{y}(t_i|\theta)\right)^2 \tag{11}$$

In MATLAB Identification Toolbox, non-linear state space models (idnlgrey) are defined and estimated using the dynamic structure as a MATLAB function. The two-tank system model (6,7) is then coded in the function *nltank* listing below.

```
function [dx,y]=nltank(t,x,u,A1,A2,k1,k2,Vmax,v0,varargin)
y = x(2);
u = (u(1)-v0)/(Vmax-v0);
dx = [-k1*sqrt(x(1))*u/A1
    (k1* sqrt(x(1))*u/A2-k2*sqrt(x(2))*u(2)/A2)];
end
```

```
sys dim=[ 1, 2, 2];
parameters=[A1, A2, Vmax, k1, k2, v0];
initstates=[h10;h20];
nlp=idnlgrey('nltank',sysdim,parameters,inistates, 0);
setpar(nlp,'Fixed',{true,true,true,false,false,false});
nlp=pem(data,nlp,'Display','Full');
k1_ est = nlp.Parameters(3).Value;
k2_ est = nlp.Parameters(4).Value;
v0_ est = nlp.Parameters(5).Value;
```

Where, the state vector x represents (h_1, h_2) and the vector input represents (V(t), a(t)), at (8). The next arguments are known parameters (A1, A2, Vmax) and unknown parameters (k1, k2, v0). To identify the order and the parameters of the model, a nonlinear grey box model object must be created following the script below.

In the script shown, sysdim=[ny,nu,nx] defines the number of outputs, inputs and states of the model, respectively. More advanced alternatives are also available, including by naming each parameter and defining its minimum and maximum value, [7]. Finally, when running the estimation procedure with (pem), k1_est, k2_est, and v0_est are the resulting estimated parameters.

The basic workflow of the identification consists in the construction of a model for the dynamics of the system *(nltank)*, setting the known and unknown variables *(nlp)* and then the acquisition of experimental data, which will be used to estimate *(pem)* the actual model values.

4. Adaptive Control

Given the model of the Water Level System in (6), (7), we can set the output $\tilde{h}_2 = h_2 - h_2^r$ as our objective function to drive to zero. In order to construct a reduced order system, we need to investigate the relative degree of the system corresponding to the desired output $\tilde{h}_2$.

$$\dot{\tilde{h}}_2 = \dot{h}_2 = -\frac{k_2\sqrt{h_2}}{A_2}a + \frac{k_1\sqrt{h_1}}{A_2}u \tag{12}$$

It can be observed that the output is relative degree one ($\rho = 1$) since $\dot{\tilde{h}}_2$ contains the control signal u.

4.1. Adaptive surface

The construction of the surface s can be formulated similar to the case of sliding mode control [4,8], considering that it can have the structure in (13), for which we propose $s = \lambda + \tilde{h}_2 + \dot{\tilde{h}}_2$ as a sliding surface. Note that $s = 0$ implies $\tilde{h}_2 \to 0$ when $\lambda \to 0$.

$$s = \left(\lambda + \frac{d}{dt}u\right)^{\rho-1} \tag{13}$$

4.2. Stability

In order to guarantee stability, we will propose a Lyapunov-like function that uses the vector of unknown parameters, p.

$$V = \frac{1}{2}s^2 + \frac{1}{2}\tilde{p}^T P^{-1}\tilde{p} \tag{14}$$

Where $\hat{p}$ is the estimated parameter vector and $\tilde{p} = \hat{p} - p$ is the estimation error. Therefore, P^{-1} is a symmetric positive definite function. By the Barbalat lemma, if $V(t, x)$ is a lower bounded continuously uniform scalar function, with $\dot{V}$ negative semidefinite along $\dot{x} = f(t, x)$, then $\dot{V}(t, x) \to 0$ as $t \to 0$.

$$\dot{V} = s\dot{s} + \dot{\tilde{p}}^T P^{-1}\tilde{p} \tag{15}$$

$$\dot{V} = s(-\frac{k_2\sqrt{h_2}}{A_2}a + \frac{k_1\sqrt{h_1}}{A_2}u) + \dot{\tilde{p}}^T P^{-1}\tilde{p} \tag{16}$$

$$\dot{V} = s\frac{k_1\sqrt{h_1}}{A_2}\left(-\frac{k_2\sqrt{h_2}}{k_1\sqrt{h_1}}a + u\right) + \dot{\tilde{p}}^T P^{-1}\tilde{p} \tag{17}$$

By the physics governing the system, it is straightforward to note that (k_1, h_1, h_2) are always positive.

4.3. Controller construction

A vector of known functions (Y) is introduced, together with the vector of unknown parameters (p).

$$Y = \frac{\sqrt{h_2}}{k_1\sqrt{h_1}}, \quad p = k_2 a \tag{18}$$

In order to drive s to zero, a controller dependent on Y and p is proposed together with a linear action on the surface s:

$$u = Y\hat{p} - ks \tag{19}$$

$$\dot{V} = s\frac{k_1\sqrt{h_1}}{A_2}(Y\tilde{p} - ks) + \dot{\tilde{p}}^T P^{-1}\tilde{p} \tag{20}$$

To obtain a negative definite $\dot{V} = \beta s^2$, $\beta > 0$, the rest of the terms will be cancelled through the selection of an adequate $\dot{\hat{p}}$. Note that $\dot{\tilde{p}} = \dot{\hat{p}}$.

$$s\frac{k_1\sqrt{h_1}}{A_2}Y + \dot{\hat{p}}^T P^{-1} = 0 \tag{21}$$

$$\dot{\hat{p}} = \frac{sk_1\sqrt{h_1}}{A_2}P^T Y^T \tag{22}$$

The adaptation law depends only in known parameters and the system states; thus, it can be implemented together with the controller. As the algorithm learns the parameter $p = k_2 a$, the control action improves and drives the surface s to zero.

IOP Conf. Series: Materials Science and Engineering **707** (2019) 012003 doi:10.1088/1757-899X/707/1/012003

5. Results

Grey Box Identification: Considering the system (7) defined with $a = 0$ we can apply ID Grey to estimate the values of both k_1, v_0, without worrying about k_2. Capturing data from our implementation for different values for V(t), and using the identification algorithm we achieve an accuracy of: 93.6% and a final prediction error (FPE) of 9.60637e-06. The model correlation is observed in figure 2(a). The resulting parameters obtained through the identification are $k_1 = 2.18 \times 10^{-4}$, $v_0 = 5.7009$. Similarly, using a known value of a, for instance when it is completely open ($a = 1$) the estimation of the parameter k_2 has an accuracy of: 98.54% and an FPE: 1.4067e-06. The model correlation is observed in figure 2(b). The resulting value for k_2 is $k_2 = 1.7286 \times 10^{-4}$. It should be noted that accuracy of the identified model has a low prediction error (FPE), however, changes on parameters can still affect the plant dynamics. It has been observed during experimentation that the constant k_2 can also change depending on the operating point of the system (dependent of the presence of laminar or turbulent flow).

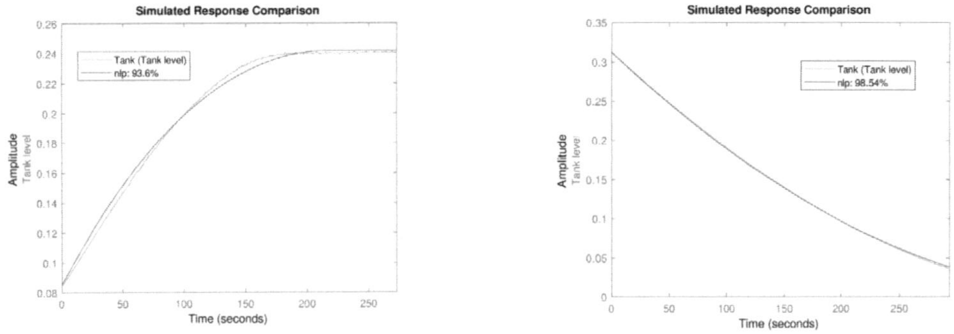

(a) (b)

Figure 2. Estimation using $a = 0$ and $a = 1$.

Setpoint tracking: A setpoint is defined at $h_2 = 0.20m$ with initial condition of the system $h_1 = 0.2149$m, $h_2 = 0.1187$m, and $a_0 = 0.2$.

Applying the adaptive control algorithm reveals that experimental data and simulation data are very similar in their evolution. Tracking results in the system closed loop system shows that the controller drives the height of tank 2 to the reference value as shown in figure 3(a). Furthermore, it is observed that the unknown parameters converge to their true value. The adaptation law has an overshoot due to the slowly evolution of the level of the tank 2. Although, it has not been formally proved, a rich reference that captures different operating points will provide an asymptotic estimation of parameters p, this behavior is observed by both simulation and experimentation.

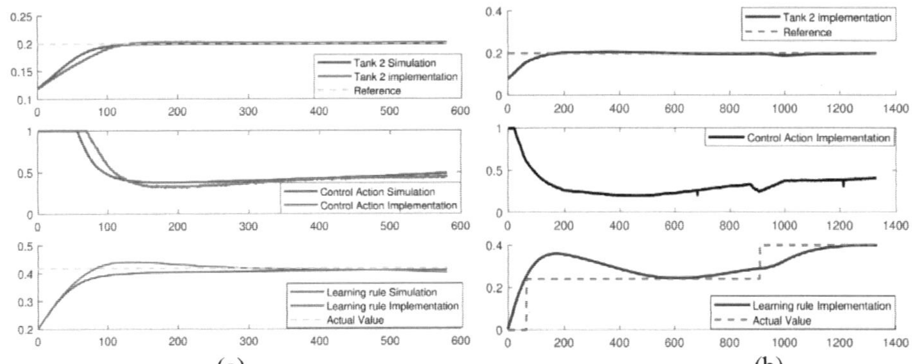

(a) (b)

Figure 3. Height variation in Tank 2, control action u and parameter estimation for (a) simulation and experiments and (b) experiments with heavy perturbations.

External disturbances: The system is tested against external disturbances when the valve 2 is operated twice during the experiment. The results in figure 3(b) indicates again a great correlation between simulation and experimentation. Also, we observe that asymptotic tracking is achieved even when disturbances are applied. Finally, the parameter a is estimated to its actual value even in the presence of two external disturbances.

6. Conclusions

We propose a workflow to work with Water Level Systems with unknown parameters by using a grey box model and an adaptive control scheme that can handle varying parameters and drive the system to the reference values. From our results it is clear that we obtain asymptotically stability of the system even in the presence of external disturbances. The system also has the power of learning the actual value of a, important for industrial operators that might be worried about open valves. Also, our controller has a very simple structure that can be easily implemented on micro-controllers as in our setup with Arduino Due with a loop frequency at 1Hz.

7. References

[1] Illes, Cosmina, Gabriel Nicolae Popa, and Ioan Filip. "Water level control system using PLC and wireless sensors." 2013 IEEE 9th International Conference on Computational Cybernetics (ICCC). IEEE, 2013.

[2] Suppachai H., Sasithorn Ch., and Warin S. "Design of PLC for Water Level Control Employing Swarm Optimization-Based PID Gain Scheduling." International Conference on Control and Robots (2018).

[3] Esfandiari, Ramin S., and Bei Lu. “Modeling and analysis of dynamic systems”. CRC press, 2018.

[4] Slotine, Jean-Jacques E., and Weiping Li. Applied nonlinear control. Vol. 199. No. 1. Englewood Cliffs, NJ: Prentice hall, 1991.

[5] Chang, Fi-John, and Ya-Ting Chang. "Adaptive neuro-fuzzy inference system for prediction of water level in reservoir." Advances in water resources 29.1 (2006): 1-10.

[6] Munasinghe, Sudath R., Min-Soeng Kim, and Ju-Jang Lee. "Adaptive neurofuzzy controller to regulate UTSG water level in nuclear power plants." IEEE Transactions on Nuclear Science 52.1 (2005): 421-429.

[7] Marvin L. "System Identification with MATLAB, Non Linear Models, ODEs and Time Series." CreateSpace Independent Publishing Platform, USA ©2018 ISBN: 9781539692317

[8] Hassan, Khalil. "Khalil, Nonlinear systems", Third Edition, Pearson New International Edition, (2014).

IOP Conf. Series: Materials Science and Engineering **707** (2019) 012004 doi:10.1088/1757-899X/707/1/012004

LQG based control design of electric vehicle driveline

Zeeshan Ali Akbar[1], Dr Fahad Mumtaz[1], Hameed Ullah[1] and Muhammad Asad Bashir[1]

[1]National University of Sciences and Technology, DEE CE&ME, Islamabad, Pakistan

Abstract. In this paper LQG (Linear-Quadratic-Gaussian) based control design for drive line of Electric vehicle is presented. Driveline oscillations is a phenomenon of fluctuating vehicle speed caused by torsional vibrations in the vehicle drive line which results as jerks in the vehicle motion. These oscillations result in uncomfortable vehicle ride which is contemporary research problem. The control objective is to achieve steady state driveline in lesser time for this purpose LQG based control is pursued with certain modification in vehicle dynamic model. This paper includes driveline modelling, control design and verification by simulations. Results of simulation show the considerable mitigation of oscillations and LQG based control for driveline of electric vehicle works very well.

1. Introduction

Instead of using precise but expensive components in drive line to remove jerks in vehicle motion we will apply control such that these jerks in vehicle motion remove. In [1] torsional oscillations are due to of fast response from the electric motor, poor damping in the driveshaft material, elasticity of various components in drive shaft, perturbation coming from road, non linearity in drive line and variable load applied to wheel. In [2] torsional oscillations lead to many problems such as driver comfort can be affected, oscillations in the torque can cause wheel blockage or slip and oscillations in the drive shaft torque will result in accelerated fatigue of mechanical components. Another problem in drive line is backlash [3]. Basically backlash is defined as play between adjacent move able parts. Sources of backlash are gear and joints of shaft. Normal load on each axle can be determined from free body diagram of static vehicle inclined to road [1,4].Where l_f, l_r represent longitudinal distance of centre of gravity from front and rear wheel

$$F_{nf} = \frac{mg(l_r \cos(\beta) + h\sin(\beta))}{l_r + l_f} - \frac{mah}{l_r + l_f} - \frac{F_D h_D}{l_r + l_f} \tag{1}$$

$$F_{nr} = \frac{mg(l_f \cos(\beta) - h\sin(\beta))}{l_r + l_f} + \frac{mah}{l_r + l_f} + \frac{F_D h_D}{l_r + l_f} \tag{2}$$

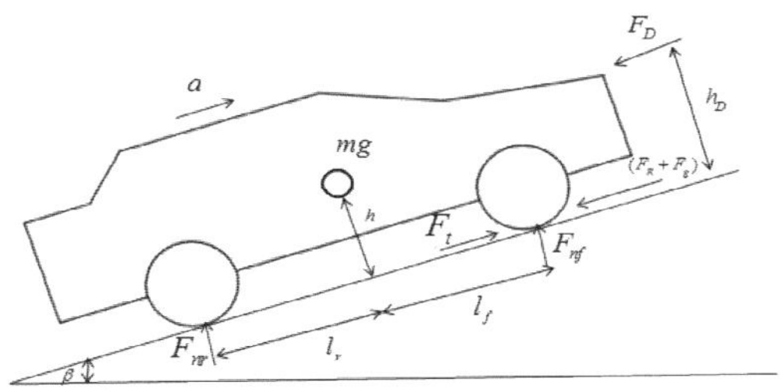

Figure1. Free Body Diagram of accelerating electric vehicle

Vehicle longitudinal dynamics is affected by driving resistance. These driving resistances consist of rolling resistance, air resistance, gradient resistance [5,6,7].Rolling resistance is roughly proportional to normal force on each axel. f is rolling resistance coefficient.

$$F_{Rr} = fF_{nr} \tag{3}$$

$$F_{Rf} = fF_{nf} \tag{4}$$

F_{Rf} , F_{Rr} are rolling resistances of front and rear wheel. Air resistance is product of dynamic pressure, frontal area of vehicle A and drag coefficient c_d .

$$F_D = \frac{1}{2}\rho c_d A v^2 \tag{5}$$

Gradient resistance is force due to of gravitation acting on vehicle body moving up or down hill.

$$F_g = mg\sin(\beta) \tag{6}$$

Where β is angle of inclination of road on which vehicle is moving. In [5,6,7] traction forces, (F_{tf}, F_{tr}) for front and rear wheel are forces acting on wheels due to contact with ground .

$$F_t = (F_{tf} + F_{tr}) = m\dot{v} + F_{drain} + F_{rolling} + F_{gradient} \tag{7}$$

In [5,8] drive line control of internal combustion engine vehicle is simulated whose system was consist of engine, drive shaft, fly wheel, clutch, wheel and transmission but we will control drive line for electric vehicle which is consist of motor, gear, Power train housing (PTH), drive shaft, backlash and wheel. PTH and drive shaft can be expressed by spring and damper in parallel. In [2] PID control and in [9] output feedback control used to compensate oscillations but we will use optimal control. In this paper simulink model is generated for drive line modelling and all forces acting on vehicle body. The effect of nonlinearity due to of backlash on drive line is analyzed. Kalman filter is used for state estimation instead of expensive sensors to measure system states. At the end LQG controller is used to remove jerks in vehicle motion for smooth drive.

The paper is organized as follow: Section 2 describes the drive line model .Section 3 describes mathematical modelling. Section 4 describes estimation of states. Section 5 describes control design and at the end section 6 and 7 describes results and conclusion.

2. Drive Line Model

Drive line model consists of electric motor, gear box, PTH, backlash, drive shaft, and wheel.

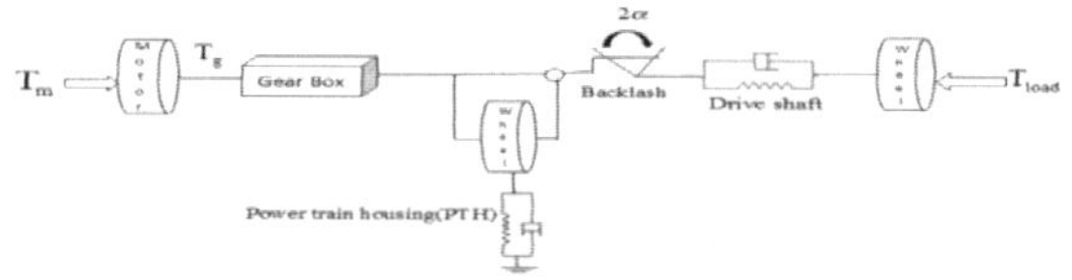

Figure2. Drive Line model

Driveshaft will deliver torque to wheels, according to the driver's torque request. It is described as spring and damper in parallel. Basically spring and damper are used to represent flexible nature of drive shaft and bushing. In [4] driveshaft can be described as

$$T_g = k_s(\theta_{g,eff} - \theta_w) + c_s(\dot{\theta}_{g,eff} - \dot{\theta}_w) \tag{8}$$

k_s is driveshaft stiffness and c_s is driveshaft damping coefficient. θ_g is shaft position connected to gear end. θ_w is shaft position connected to wheel end. Power train housing (PTH) consists of non rotating parts of electric motor and gear box mounted together [4]. Where k_{pt} is stiffness coefficient, c_{pt} is damping coefficient and j_{pt} rotational inertia of PTH. Rotational acceleration of lumped PTH is

$$\dot{w}_{pt} = \frac{T_g - K_{pt}\theta_{pt} - C_{pt}w_{pt}}{j_{pt}} \tag{9}$$

$$\text{Effective rotational acceleration} = \dot{w}_{g,eff} = \dot{w}_g - \dot{w}_{pt} \tag{10}$$

3. Mathematical Modelling

3.1. Wheel acceleration

In [4,5] according to Newton's second law of rotational motion

$$J_w \dot{w}_w = T_g - T_t \tag{11}$$

Where T_t is traction torque which is related to traction force F_t . T_g is drive shaft torque. J_w is wheel inertia. $\dot{w}_w$ is wheel acceleration. F_{res} is sum of rolling, air and gradient resistances.

$$J_w \dot{w}_w = T_g - (ma + F_{res})r \tag{12}$$

$$\dot{w}_w = \frac{T_g - T_{res}}{J_w + mr^2} \tag{13}$$

$$\dot{w}_w = \frac{k_s(\theta_g - \theta_w)}{(J_w + mr^2)} + \frac{c_s w_m}{(J_w + mr^2)i} - \frac{c_s w_w}{(J_w + mr^2)} - \frac{T_{res}}{(J_w + mr^2)} \tag{14}$$

3.2. Torsion

Torsion in drive shaft is difference between shaft position connected to gear end θ_g and shaft position connected to wheel end θ_w.

3.3. Motor acceleration

In [1,4] motor and transmission dynamics are given as

$$J_d \dot{w}_m = T_m - \frac{T_g}{i} \tag{15}$$

Where J_m is motor inertia, $\dot{w}_m$ is motor rotational acceleration, T_m is input torque of motor, i is final gear ratio, i_{g1} is primary gear ratio, i_{g2} is secondary gear ratio, J_1, J_2, J_3 are primary secondary and final gear ratio. Total speed conversion from motor to gear box out coming shaft is $w_g = \frac{w_m}{i}$

J_d =Drive Train Inertia $= J_m + \frac{J_t}{i^2}$, J_t = Inertia of Transmission$= J_1 i^2 + J_2 i_{g2}^2 + J_3$

$$\dot{w}_m = \frac{T_m}{J_d} - \frac{T_g}{J_d i} \tag{16}$$

$$\dot{w}_m = \frac{-k_s(\theta_g - \theta_w)}{j_d i} - \frac{c_s w_m}{j_d i^2} + \frac{c_s w_w}{j_d i} + \frac{T_m}{j_d} \tag{17}$$

3.4 State Space Model

State space model of drive line can be represent as $x = [\theta_g - \theta_w \quad w_m \quad w_w]^T$

$$\dot{\vec{x}} = A\vec{x} + B\vec{u} + E\vec{u}_d = \begin{bmatrix} 0 & \frac{1}{i} & -1 \\ \frac{-k_s}{j_d i} & -\frac{c_s}{j_d i^2} & \frac{c_s}{j_d i} \\ \frac{k_s}{(J_w + mr^2)} & \frac{c_s}{(J_w + mr^2)i} & -\frac{c_s}{(J_w + mr^2)} \end{bmatrix} \vec{x} + \begin{bmatrix} 0 \\ \frac{1}{j_d} \\ 0 \end{bmatrix} T_m - \begin{bmatrix} 0 \\ 0 \\ \frac{r}{(J_w + mr^2)} \end{bmatrix} F_{res} \tag{18}$$

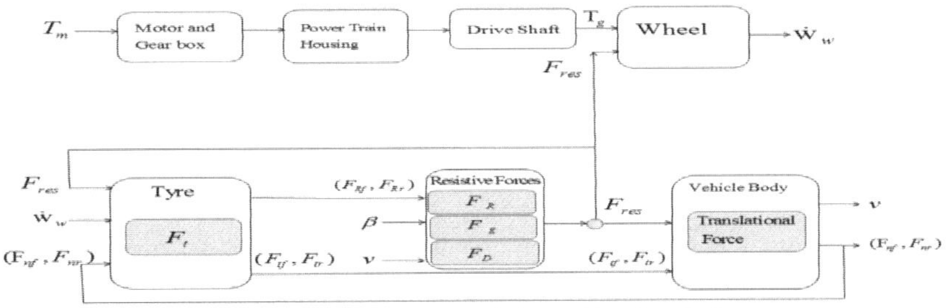

Figure3. Drive Line Model used in simulation

4. Estimator

Generally it is not possible to measure all states of system because of sensors used to measure states of system are very expensive and measured states of system using these sensors are too noisy. To solve these problems we estimate unknown states of system with the help of estimator using mathematical model of system and state measurement. Plant with process noise w and measurement noise v is

$$\dot{x} = Ax + Bu + w \quad , \quad y = Cx + v \tag{19}$$

Covariance for noises $$Q = E(ww^T), \qquad R = E(vv^T) \tag{20}$$

Kalman filter equation $$\hat{x}_k = A\hat{x}_{k-1} + Bu_k + K_k[y_k - C(A\hat{x}_{k-1} + Bu_k)] \tag{21}$$

For drive line of electric vehicle wheel speed and motor speed both can measure but wheel speed measurement is far lower than motor speed measurement so we will use motor speed as measurement state for estimator so $y = C\hat{x} = \begin{bmatrix} 0 & 1 & 0 \end{bmatrix}\hat{x}$

In [10] Kalman filter uses system model and state measurement to find kalman gain K_k minimizing error covariance P_k^-. Kalman estimator is a two step process

Prediction: Update:

$$\begin{aligned} &\hat{x}_k^- = A\hat{x}_{k-1} + Bu_k & &K_k = (P_k^- C^T)/(CP_k^- C^T + R) \\ &P_k^- = AP_{k-1}A^T + Q & &\hat{x}_k = \hat{x}_k^- + K_k(y_k - C\hat{x}_k^-) \\ & & &P_k = (I - K_k C)P_k^- \end{aligned} \tag{22}$$

5. Control Design

To bring the drive line to steady state quickly for smoother drive, controller regulates torque demand to electric motor.LQG is combination of kalman filter and linear quadratic regulator (LQR). Kalman estimates unknown states of system and LQR regulates torque demand to electric motor. In [11] the state control will try to minimize cost function J

$$J = \frac{1}{2}\int_0^\infty (x^T Q_{lqr} x + u^T R_{lqr} u)dt \tag{23}$$

Q_{lqr} and R_{lqr} are weighted squares of deviation of states from target and weighted squares of control activity. Square are used because these lead to easier analysis and well behaved solutions. Q_{lqr} and R_{lqr} allow the trade of between input activity and rate of convergence. Term $x^T Q_{lqr} x$ is related to convergence rate such as Rise time, Settling time. Term $u^T R_{lqr} u$ is related to aggressive use of input. P_{lqr} is a positive definite matrix.

$$u = -K_{lqr} x \tag{24}$$

$$K_{lqr} = R_{lqr}^{-1} B^T P_{lqr} \tag{25}$$

$$A^T P_{lqr} + P_{lqr} A - P_{lqr} B R_{lqr}^{-1} B^T P_{lqr} + Q_{lqr} = 0 \tag{26}$$

$$\dot{x} = (A - BK_{lqr})x = (A - BR_{lqr}^{-1} B^T P_{lqr})x \tag{27}$$

6. Results

We can see oscillations in drive shaft and electric vehicle acceleration for step input of 200 Nm motor torque in figures [4-6].Torque oscillations in drive shaft relate to oscillations in acceleration of electric vehicle. Controlled response of drive shaft and electric vehicle is shown in figures [5-9].

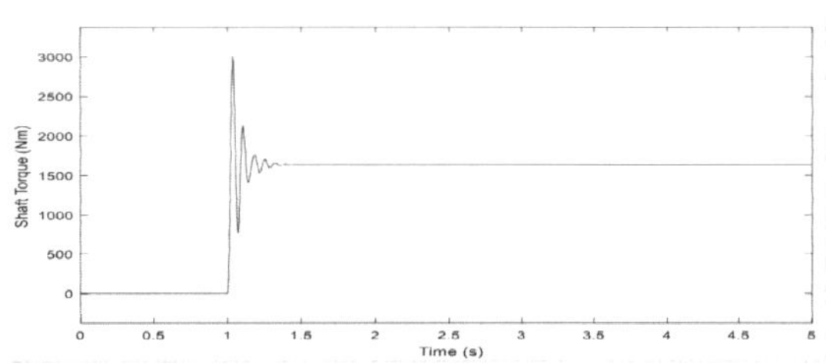

Figure 4. Shaft torque for 200 Nm step input

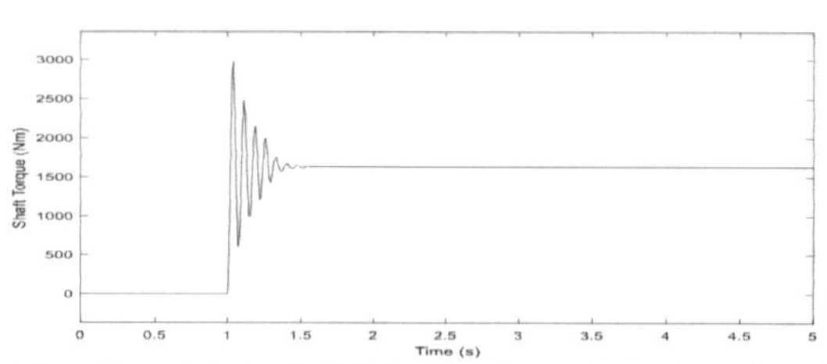

Figure 5. Shaft torque with 30 degree backlash

IOP Conf. Series: Materials Science and Engineering **707** (2019) 012004 doi:10.1088/1757-899X/707/1/012004

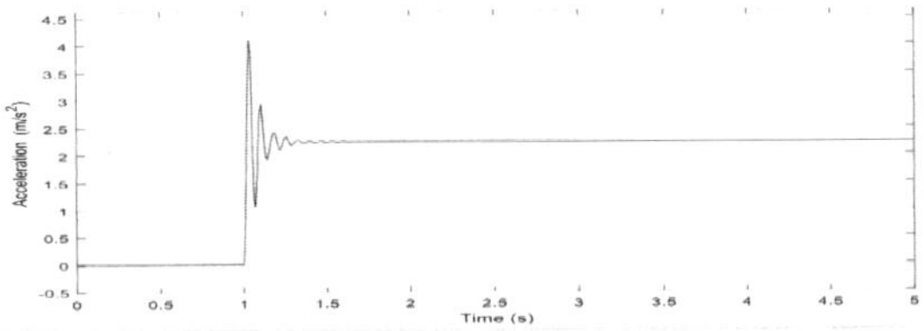

Figure 6. Vehicle acceleration for 200 Nm step input

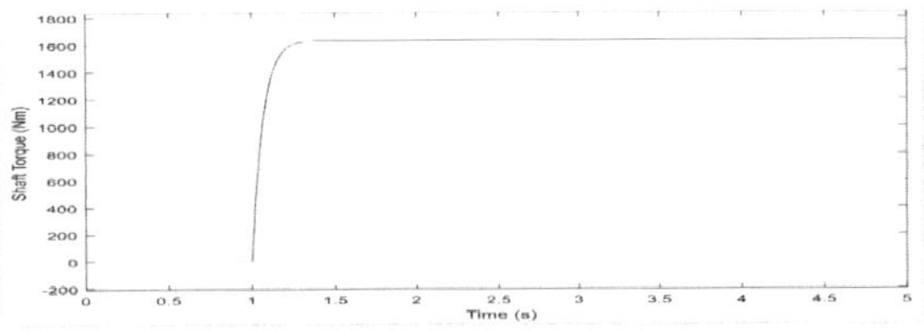

Figure 7. Shaft torque with LQG

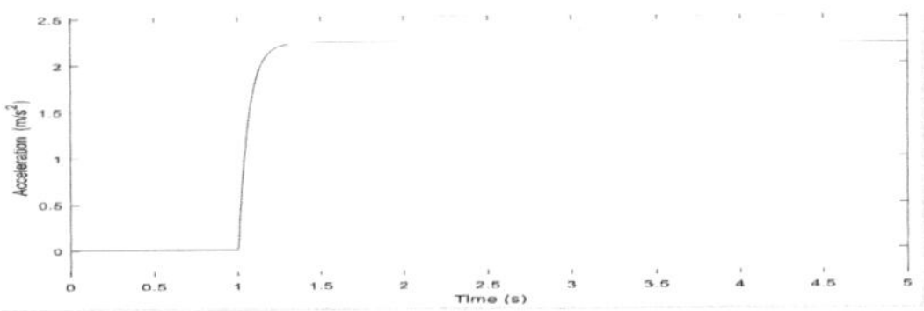

Figure 8. Vehicle acceleration with LQG

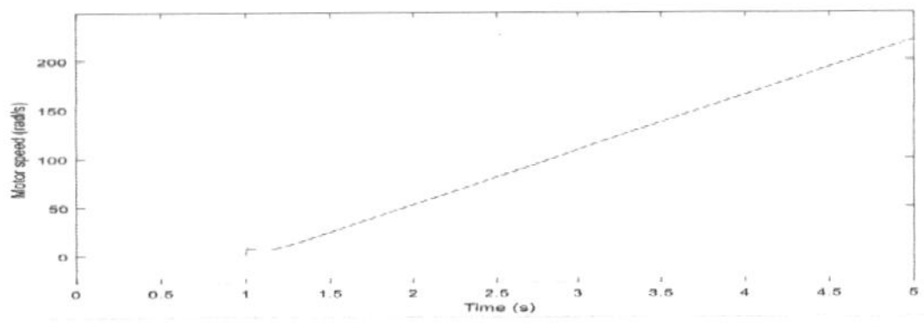

Figure 9. Motor speed with LQG

7. Conclusion

In this paper we have presented the concept of drive line control and how it can be applied to automotive application. Due to torsional vibrations vehicle motion results as jerks in start so an anti jerk controller has been designed for an electric vehicle using LQG to meet multiple objectives of velocity tracking performance and enhance drivability. We use Kaman filter for estimation of states using motor speed as measurement state. LQG control is a feasible and successful in reducing vehicle oscillations which regulates torque demand to electric motor. We have also implemented dynamics of wheel, tyre, resistive forces and vehicle body in this paper. The control algorithm is implemented in existing motor control without any additional cost.

8. References

[1] Batra, Mohit. "Dynamics and Model-Predictive Anti-Jerk Control of Connected Electric Vehicles." (2018).

[2] Rodríguez, José Manuel, Rubén Meneses, and Javier Orús. "Active vibration control for electric vehicle compliant drivetrains." IECON 2013-39th Annual Conference of the IEEE Industrial Electronics Society. IEEE, 2013.

[3] Sainio, Joonas. "Backlash compensation in electric vehicle powertrain." (2016).

[4] HERMANSSON, VICTOR, and KEDARNATH MOPARTHI. "Control of an Electric Vehicle Powertrain to Mitigate Shunt and Shuffle."

[5] Northcote, Nicholas M. The modelling and control of an automotive drivetrain. Diss. Stellenbosch: University of Stellenbosch, 2006.

[6] George, Shelma, Renji V. Chacho, and K. Salitha. "Modelling and simulation of Electric Vehicle power train in SEQUEL." 2014 IEEE International Conference on Power Electronics, Drives and Energy Systems (PEDES). IEEE, 2014.

[7] Murillo, Victor Hugo González, C. Perez-Rojas, and SigridtGarcíaMartínez. "Modeling and simulation of the power train of an electric vehicle." 2013 IEEE International Autumn Meeting on Power Electronics and Computing (ROPEC). IEEE, 2013.

[8] Fredriksson, Jonas, Henrik Weiefors, and Bo Egardt. "Powertrain control for active damping of driveline oscillations." Vehicle system dynamics 37.5 (2002): 359-376.

[9] Mo, C. Y., A. J. Beaumount, and N. N. Powell. Active control of driveability. No. 960046. SAE Technical Paper, 1996.

[10] Li, Qiang, et al. "Kalman filter and its application." 2015 8th International Conference on Intelligent Networks and Intelligent Systems (ICINIS). IEEE, 2015.

[11] Fredriksson, Jonas. "Improved driveability of a hybrid electric vehicle using powertrain control." International Journal of Alternative Propulsion 1.1 (2006): 97-111.

IOP Conf. Series: Materials Science and Engineering **707** (2019) 012005 doi:10.1088/1757-899X/707/1/012005

Sampled-data control of pan-tilt platform using discrete-time high gain observer

Hameed Ullah[1], **Fahad Mumtaz Malik**[1], **Anjum Saeed**[1], **Zeeshan Ali Akbar**[1] **and Sajjad Hussain**[1]

[1]National University of Sciences and Technology, DEE CE&ME, Islamabad, Pakistan

Abstract. In this paper we have presented the design of sampled-data control algorithm for Pan-tilt platform using Discrete-time High Gain Observer (DHGO). The Pan-Tilt platform is a 2-axis serial robotic manipulator having a nonlinear dynamic model. The design of the control algorithm is carried out in 2 steps. In 1st steps, a continuous time state feedback control law based on feedback linearization is designed which is subsequently discretized. In the second stage, the states of the system are estimated using DHGO. To evaluate the performance of the suggested algorithm extensive simulations are carried out in Matlab Simulink.

1. Introduction

Pan-tilt platform (PTP) is a complex combination of electro-mechanical components used to carry high-precision devices such as camera or range finding sensors. The Schematic of a Pan-tilt platform has 2 degrees of freedom (DOF) and is composed of a base, a rotatable pan mechanism and a tilt mounted on the pan to orient the device in the desired direction. Pan-tilt platform also known as a gamble has potential applications in inspections, monitoring, unmanned systems and so on [1]. In all these applications, the gimbal is required to follow a profile of the pan and tilt angles termed as the reference angles/ trajectories. However, because of the presence of nonlinearities in the system dynamic model and other parametric uncertainties, achieving high precision stabilization or trajectory tracking is difficult. Furthermore, these systems are prone to external disturbances such as high-frequency vibrations, wind, and sudden shocks etc., thus a sophisticated control algorithm, which will ensure the desired performance specification irrespective of the system model parametric uncertainties and external disturbances, is required.

The solution turns out to be output feedback linearization based control algorithm. Output feedback linearization, which converts non-linear systems into linear systems, is a powerful technique to achieve the desired performance specification. However, to implement this control algorithm all the system parameters and states needs to be known, which practically is not possible. To solve this problem high gain observer (HGO) has been extensively used as an essential tools for the design of feedback control for non-linear system [2]. In literature two approaches have been reported for the implementation of the sampled data control algorithm. In the first approach, the control algorithm in continuous time is designed by means of the plant continuous-time model after that it is discretized and implemented while in the 2nd approach the controller is designed in the discrete-time by means of the system discrete time model [3]. In this study, we will opt for the first approach.

This work presents the design of a sampled data HGO based on the feedback controls for 2DOF pan-tilt platform. To achieve the aforementioned gaol first a dynamic model for the system is presented. Then a continuous time control algorithm is defined for the system using a continuous time

IOP Conf. Series: Materials Science and Engineering **707** (2019) 012005 doi:10.1088/1757-899X/707/1/012005

model. Using Forward Difference (FD) method the continuous time control algorithm is discretized and is then implemented using the 2 DOF model of the gamble.

The paper is organized as, the mathematical modelling of 2DOF system is explained in section 2, followed by the design of continuous-time feedback linearization based control and HGO in section 3 and 4 respectively. The discretization of continuous time control algorithm is discussed in section 5. While in section 6 results and in section 7 conclusion are discussed.

2. Mathematical modelling

The mathematical modelling of moving objects is accomplished by attaching frames to its links. The frame assignment for our system is as shown in fig. 1. Frame 0, also known as the base frame, in case of gamble it is attached to the unmanned aerial vehicle (UAV). The motion of first rotary joint represented by frame 1. Frame 2 is attached to the second joint. It is important to note that the rotation of the 1st link is about z-axis by an angle " θ_1" as known as pan angle and the actuator torque " τ_1"

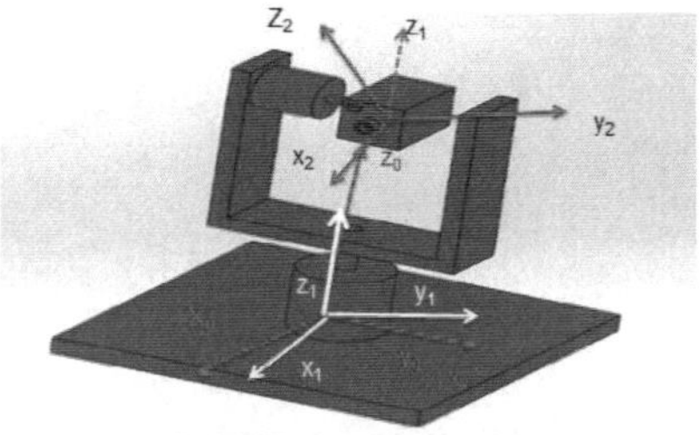

Figure 1. Schematic model with frame assignments of a 2 DOF Gimbal

cause this rotation. So, the dynamic equation for the first joint mathematically given as [4]-[6],

$$\ddot{\theta}_1 = \frac{-2I_{2x}\dot{\theta}_1\dot{\theta}_2\cos\theta_2\sin\theta_2+2I_{2z}\dot{\theta}_1\dot{\theta}_2\cos\theta_2\sin\theta_2+\tau_1}{(I_{2x}\sin^2\theta_2+I_{2z}\cos^2\theta_2+I_{1z})} \tag{1}$$

Where $I_{2x}$and I_{2z} are the moments of inertia for link-2 about x and z-axis respectively and I_{1z} is the moment of inertia for link-1 about z-axis.

While the rotation of the 2nd link is about y-axis by an angle " θ_2" known as tilt angle and the actuator toque " τ_2" is responsible for the rotation of the second joint. So, the dynamic equation for the second joint mathematically given as [4]-[6],

$$\ddot{\theta}_2 = \frac{I_{2x}\dot{\theta}_1^2\cos\theta_2\sin\theta_2-I_{2z}\dot{\theta}_1^2\cos\theta_2\sin\theta_2+\tau_2}{I_{2y}} \tag{2}$$

Where $I_{2y}$is the moment of inertia for link-2 about y-axis.

3. Feedback linearization of 2-DOF gimbal

Feedback linearization is one of the most important and widely used techniques for controlling nonlinear system. The basic idea of feedback linearization (FBL) is to choose appropriate control inputs which will lineariz the nonlinear system by cancelling out the nonlinearities of the system, so that feasible linear control techniques could also be used for these systems [7]. The 1st stage for feedback linearization is to write the model of the system in state space form. To generate state-space form of nonlinear system explained in equation (1) and (2), we will consider the joint angles as the state variable and the motor torques as the input to the system, given as follow,

$$x_1 = \theta_1, x_3 = \theta_2, u_1 = \tau_1, u_2 = \tau_2$$

Substituting these values in equation (1) and (2), results in the state space model for gamble, given as

$$\dot{x}_1 = x_2 \,,\, \dot{x}_2 = \frac{-2I_{2x}x_2x_4\sin x_3\cos x_3+2I_{2z}x_2x_4\sin x_3\cos x_3+u_1}{I_{2x}\sin^2 x_3+I_{2z}\cos^2 x_3+I_{1z}},$$

$$\dot{x}_3 = x_4 \,,\, \dot{x}_4 = \frac{I_{2x}x_2^2\sin x_3\cos x_3-I_{2z}x_2^2\sin x_3\cos x_3+u_2}{I_{2y}} \tag{3}$$

3.1. Controller Design

In compact form equation (3) can be written as,

$$\dot{x}_1 = x_2 \,,\, \dot{x}_2 = f_1(x) + g_1(x)u_1 \,,\, \dot{x}_3 = x_4 \,,\, \dot{x}_4 = f_2(x) + g_2(x)u_2 \tag{4}$$

Where $f_1(x)$, $f_2(x)$, $g_1(x)$ and $g_2(x)$ are defined as follow,

$$f_1(x) = \frac{2(I_{2z}-I_{2x})x_2x_4 \sin x_3 \cos x_3}{I_{2x}\sin^2 x_3 + I_{2z}\cos^2 x_3 + I_{1z}} \,,\, f_2(x) = \frac{(I_{2x}-I_{2z})x_2^2 \sin x_3 \cos x_3}{I_{2y}},$$

$$g_1(x) = \frac{1}{I_{2x}\sin^2 x_3 + I_{2z}\cos^2 x_3 + I_{1z}} \,,\, g_2(x) = \frac{1}{I_{2y}}$$

The state feedback control input for system model based on feedback linearization approach [7] and [8] can be defined as,

$$u_1 = (-f_1(x) + v_1)g_1^{-1}(x) \,,\, u_2 = (-f_2(x) + v_2)g_2^{-1}(x) \tag{5}$$

Putting this control inputs in equation (4) will results in linear counterpart of the nonlinear system, as

$$\dot{x}_1 = x_2 \,,\, \dot{x}_2 = v_1 \,,\, \dot{x}_3 = x_4 \,,\, \dot{x}_4 = v_2 \tag{6}$$

Where v_1 and v_2 are known as the auxiliary inputs and mathematically given as $v_1 = -k_1x$ and $v_2 = -k_2x$ where the gains ($k_1$and k_2) are selected such that the closed loop system eigenvalues lies in the left half plane. The closed loop system poles are selected to be at -5, -6, -5.5 and -6.5, resulting in the final equations for v_1 and v_2 to be

$$v_1 = -35.569x_1 - 11.968x_2 - x_3 - 0.175x_4 \,,\, v_2 = -1.005x_1 - 0.175x_2 - 30.181x_3 - 11.031x_4$$

4. Continuous-time Design

Let a multiple input multiple output (MIMO) nonlinear system mathematically defined [12] as

$$\dot{x} = Ax + B\emptyset(x,u) \,,\, y = Cx \tag{7}$$

where $x \in R^n$ represents the state vector, $u \in R^r$ represents the control input whereas $y \in R^r$ is the output. Where A is the $n \times n$ matrix, B is the $n \times r$ matrix and C is the $r \times n$ matrix are given by

$$A = block\ diag[A_1,\ldots,A_r] \,,\, A_j = \begin{bmatrix} 0 & 1 & \ldots & \ldots & 0 \\ 0 & 0 & 1 & \ldots & 0 \\ \vdots & & & & \vdots \\ 0 & \ldots & \ldots & 0 & 1 \\ 0 & \ldots & \ldots & \ldots & 0 \end{bmatrix}_{n_j \times n_j} \,,\, B = block\ diag[B_1,\ldots,B_r] \,,$$

$$B_j = [0 \;\; 0 \;\; \ldots \;\; \ldots \;\; 1]^T_{n_j \times 1} \text{ and } C = block\ diag[C_1,\ldots,C_r] \,,\, C_j = [1 \;\; 0 \;\; \ldots \;\; \ldots \;\; 0]_{1 \times n_j}$$

Where $1 \leq j \leq r$ and $n = n_1 + \ldots + n_r$.

4.1. HGO Equation

The continuous time HGO equation can be written [9], [12] as

$$\dot{\hat{x}} = A\hat{x} + B\emptyset_0(\hat{x},u) + H(y - C\hat{x}) \tag{8}$$

Here $\emptyset_0(\hat{x},u)$is the nonlinear function $\emptyset(\hat{x},u)$ nominal model where $H = block\ diag[H_1,\ldots,H_r]$,

$$H_j = \begin{bmatrix} \frac{\alpha_1^j}{\varepsilon} & \frac{\alpha_2^j}{\varepsilon^2} & \ldots & \frac{\alpha_{n_j-1}^j}{\varepsilon^{n_j-1}} & \frac{\alpha_{n_j}^j}{\varepsilon^{n_j}} \end{bmatrix}^T_{n_j \times 1}$$

is known as the observer gain and ε is a positive small parameter. Whereas the α_i^j are the positive constants to be selected in a manner that the roots of

$$s^{n_j} + \alpha_1^j s^{n_j-1} + \ldots + \alpha_{n_j-1}^j s + \alpha_{n_j}^j = 0$$

have negative real parts, for all $j = 1,\ldots,r$.

5. Sampled-data Control

5.1. HGO Discretization

To avoid the inherent ill-conditioning in realization of HGO, caused by a small value of ε, we will start by scaling the observer variables [10], let $\varphi = E\hat{x}$ so

$$\dot{\varphi} = \frac{1}{\epsilon}[A_0\varphi + H_0 y + \varepsilon^n B\emptyset_0\,(E^{-1}\varphi, u)] \tag{9}$$

Where $\hat{x} = E^{-1}\varphi$, $E = diag[1, \varepsilon, \dots, \varepsilon^{n-1}]$, $A_0 = \varepsilon E(A - HC_c)E^{-1} := A - H_0C$, $EB = \varepsilon^{n-1}B$ and $H_0 = \varepsilon EH$

5.2. *Discrete-time Nonlinear HGO*

By using the FD method [10]-[11] the nonlinear HGO observer (9) can be discretised when $\emptyset_0 \neq 0$, we get

$$q(k+1) = A_d q(k) + B_d y(k) + T\varepsilon^{n-1}B\emptyset_0(E^{-1}q(k), u(k)) \tag{10}$$

Where $\hat{x}(k) = C_d q(k)$, $A_d = (I + \alpha A_0)$, $B_d = \alpha H_0$ and $C_d = E^{-1}$

In the FD method, we consider that "α" is sufficiently smaller value to make sure that the eigenvalues of $A_d = (I + \alpha A_0)$ lies inside the unit circle. Whereas in our system of 2-DOF Gimbal of equation (4) we have that

$$A = \begin{bmatrix} 0 & 1 & 0 & 0 \\ 0 & 0 & 0 & 0 \\ 0 & 0 & 0 & 1 \\ 0 & 0 & 0 & 0 \end{bmatrix}, B = \begin{bmatrix} 0 & 0 \\ 1 & 0 \\ 0 & 0 \\ 0 & 1 \end{bmatrix}, C = \begin{bmatrix} 1 & 0 & 0 & 0 \\ 0 & 0 & 1 & 0 \end{bmatrix}, \phi_o = \begin{bmatrix} f_1(x) + g_1(x)u_1 \\ f_2(x) + g_2(x)u_2 \end{bmatrix}, H = \begin{bmatrix} \frac{\alpha_1}{\varepsilon_1} & \frac{\alpha_2}{\varepsilon_1^2} & 0 & 0 \\ 0 & 0 & \frac{\alpha_3}{\varepsilon_2} & \frac{\alpha_4}{\varepsilon_2^2} \end{bmatrix}^T$$

By substituting these values of A, B , C, $\emptyset_0$ and H in equation (9), we get A_0 and H_0 which will be further used in equation (10), we get the final equations of the discrete time High Gain Observer as

$$\hat{x}_1[k+1] = \left(1 - \frac{\alpha\varepsilon\alpha_1}{\varepsilon_1}\right)\hat{x}_1[k] + \alpha\varepsilon\hat{x}_2[k] + \left(\frac{\alpha\varepsilon\alpha_1}{\varepsilon_1}\right)y_1[k]$$

$$\hat{x}_2[k+1] = \frac{1}{\varepsilon}\left[\left(-\frac{\alpha\varepsilon^2\alpha_2}{\varepsilon_1^2}\right)\hat{x}_1[k] + \varepsilon\hat{x}_2[k] + \left(\frac{\alpha\varepsilon^2\alpha_2}{\varepsilon_1^2}\right)y_1[k] + T\varepsilon^3(f_1(\hat{x}) + g_1(\hat{x})u_1)\right]$$

$$\hat{x}_3[k+1] = \frac{1}{\varepsilon^2}\left[\left(1 - \frac{\alpha\varepsilon^3\alpha_3}{\varepsilon_2}\right)\varepsilon^2\hat{x}_3[k] + \alpha\varepsilon^3\hat{x}_4[k] + \left(\frac{\alpha\varepsilon^3\alpha_3}{\varepsilon_2}\right)y_2[k]\right]$$

$$\hat{x}_4[k+1] = \frac{1}{\varepsilon^3}\left[\left(-\frac{\alpha\varepsilon^4\alpha_4}{\varepsilon_2^2}\right)\varepsilon^2\hat{x}_3[k] + \varepsilon^3\hat{x}_4[k] + \left(\frac{\alpha\varepsilon^4\alpha_4}{\varepsilon_2^2}\right)y_2[k] + T\varepsilon^3(f_2(\hat{x}) + g_2(\hat{x})u_2)\right]$$

6. Simulations and results

To evaluate the performance of the suggested control algorithms simulations were carried out in the Matlab/Simulink by applying the proposed control law to a 2 DOF pan tilt platform, discussed in section 2. A comparison of DHGO based and simple feedback control is also given which shows the efficacy of the suggested control algorithm. Simulation results are presented in Fig. 2 -10. The selected DHGO parameters for these simulations are, $\varepsilon = 0.8$, $T = 0.01$, $\alpha = 0.01961$, $\alpha_1 = 11$, $\alpha_2 = 30$ and $\alpha_3 = 12$, $\alpha_4 = 35.75$. Fig. 2 –5, shows the comparison of system states under simple state feedback control law with those estimated via DHGO. The error in estimating the states x_1 to x_4 via DHGO is shown in figure 6-9. From these figures it is evident that the suggested DHGO is capable to accurately track the system states. The designed control input plot of input u_1 and input u_2 are shown in Fig. 10.

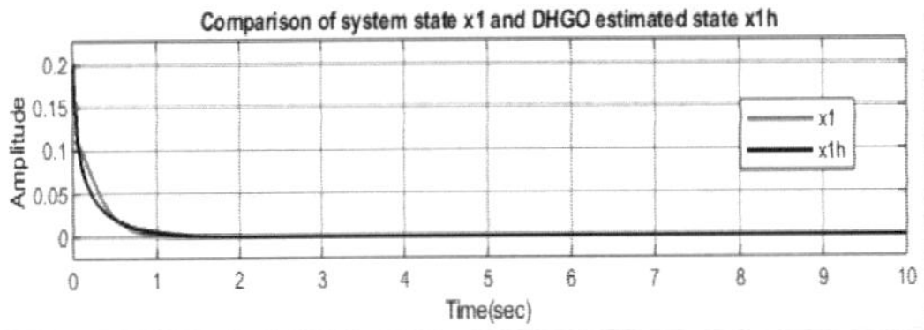

Figure 2. Comparison between system state $\boldsymbol{x_1}$ and DHGO estimated state $\hat{\boldsymbol{x}}_1$

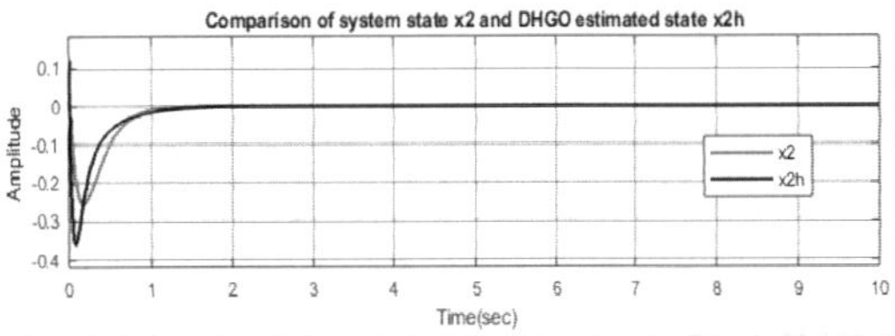

Figure 3. Comparison between system state $\boldsymbol{x_2}$ and DHGO estimated state $\hat{\boldsymbol{x}}_2$

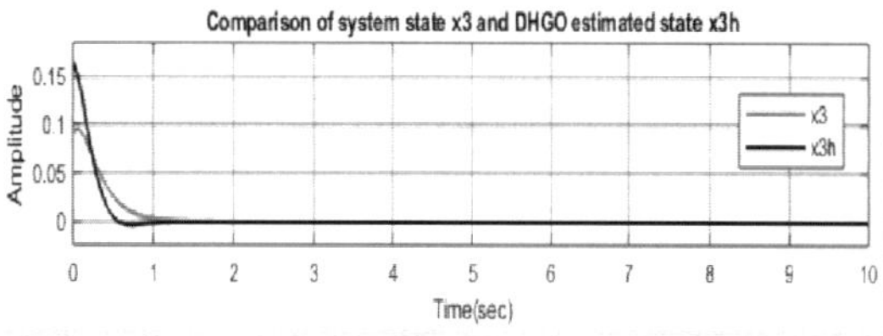

Figure 4. Comparison between system state x_3 and DHGO estimated state $\hat{x}_3$

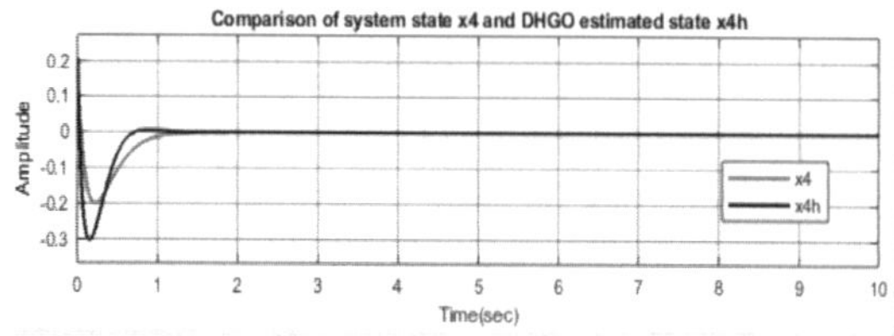

Figure 5. Comparison between system state x_4 and DHGO estimated state $\hat{x}_4$

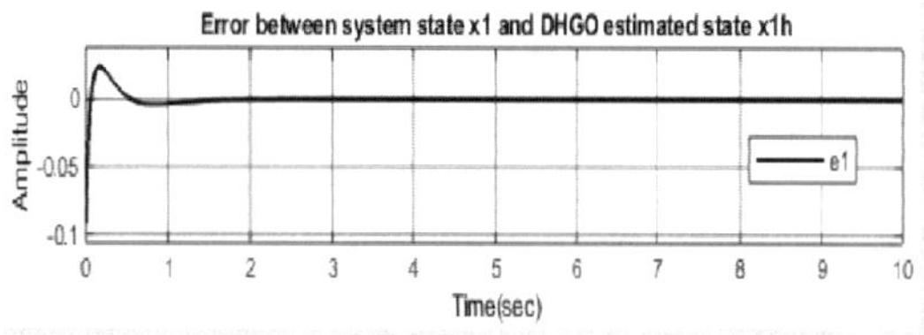

Figure 6. Error between system state x_1 and DHGO estimated state $\hat{x}_1$

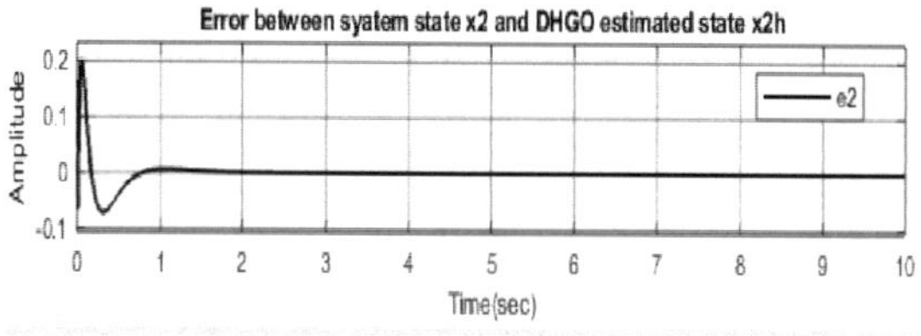

Figure 7. Error between system state x_2 and DHGO estimated state $\hat{x}_2$

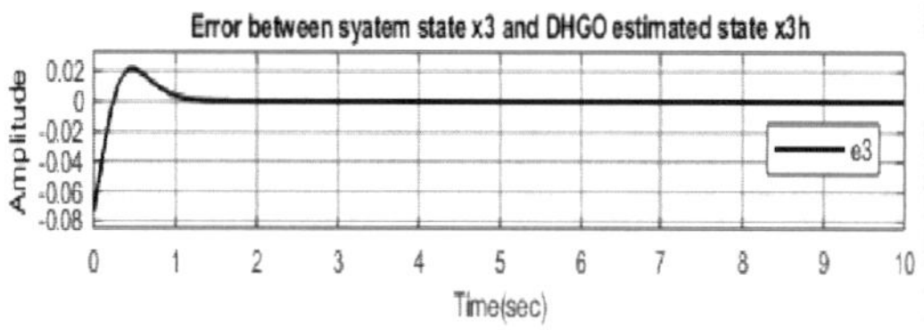

Figure 8. Error between system state x_3 and DHGO estimated state $\hat{x}_3$

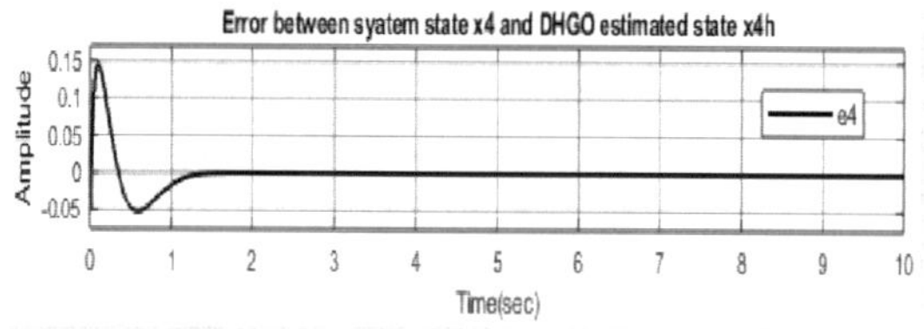

Figure 9. Error between system state x_4 and DHGO estimated state $\hat{x}_4$

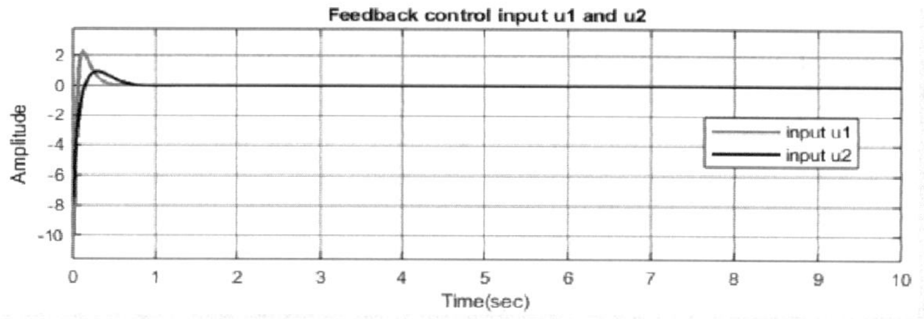

Figure 10. Feedback control input u_1 and u_2 for DHGO

7. Conclusion

In this paper, we have discussed the design of sampled data control algorithm based on DHGO for a nonlinear 2 DOF pan tilt platform used for aerial imaging. The whole control scheme is developed in two steps, in first step HGO is design in continuous time where in the 2nd step HGO is discretised by using the FD method. The convergence of the system states shows the efficacy of DHGO based feedback linearization control. In addition to the performance recovery, another advantage of the proposed control algorithm is the reduction in implementation cost achieved by the simplification of control law through feedback linearization.

8. References

[1] Ji, Yingfeng, Ronald A. Perez, and Ryoichi S. Amano. "Modeling and Control of Underwater Pan/Tilt Camera Tracking System: Geometry Modeling and Tracking Control." ASME 2009 International Mechanical Engineering Congress and Exposition. American Society of Mechanical Engineers, 2009.

[2] Khalil, Hassan K., and Laurent Praly. "High-gain observers in nonlinear feedback control." International Journal of Robust and Nonlinear Control 24.6 (2014): 993-1015.

[3] Khalil, Hassan K. "Performance recovery under output feedback sampled-data stabilization of a class of nonlinear systems." *IEEE Transactions on Automatic Control* 49.12 (2004): 2173-2184.

[4] Altaf, Omair, et al. "Extended order high gain observer based stabilization of 2 DOF pan tilt platform for aerial imaging system." *Proceedings of the 2018 3rd International Conference on Reliability Engineering (ICRE 2018).*

[5] Craig, John J. *Introduction to Robotics: Mechanics & Control*, (1986).

[6] Ahmad, Naseem, et al. "Design of Adaptive SMC for EOD Robotics Manipulator Arm.".

[7] Isidori, Alberto. *Nonlinear Control Systems*, Springer, (1985).

[8] Khalil, Hassan K., and Jessy W. Grizzle. *Nonlinear systems*, vol.**3** (2002).

[9] Ahrens, Jeffrey H., and Hassan K. Khalil. "High-gain observers in the presence of measurement noise: A switched-gain approach." Automatica 45.4 (2009): 936-943.

[10] Dabroom, Ahmed Mohamed, and Hassan K. Khalil. "Output feedback sampled-data control of nonlinear systems using high-gain observers." *IEEE Transactions on Automatic Control* 46.11 (2001): 1712-1725.

[11] Dabroom, Ahmed M., and Hassan K. Khalil. "Discrete-time implementation of high-gain observers for numerical differentiation." *International Journal of Control* 72.17 (1999): 1523-1537.

[12] Atassi, Ahmad N., and Hassan K. Khalil. "A separation principle for the stabilization of a class of nonlinear systems." 1997 European Control Conference (ECC). IEEE, 1997.

IOP Conf. Series: Materials Science and Engineering **707** (2019) 012006 doi:10.1088/1757-899X/707/1/012006

Simple discrete time traction control for a two-wheeled mobile robot

M Dahlan[1], A Sutapun[1] and V Sangveraphunsiri[1]

[1]Chulalongkorn University, Department of Mechanical Engineering, 254 Phayathai, Wang Mai, Pathum Wan, Bangkok, 10330, Thailand

Abstract. Traction controls usually need dynamic parameters as an approximated reference in the model, such as inertia and friction or heavily computational estimator, such as slip ratio estimator and friction estimator. This paper is shown that the dynamic parameters in the traction controller can be negligible in some cases, internal sensor state variables and gain parameters are enough to retain some traction controllability. The controller is derived from the model following control method (MFC) on a discrete domain. Experiments are performed, in order to evaluate the efficiency of the simplified controller using a two-wheeled mobile robot, by comparing slippage between the position and velocity control scheme with and without the traction control scheme. The experimental results show that the proposed traction control scheme is able to suppress the slip and very simple to implement.

1. Introduction

Nowadays, most commercialized mobile robots need to be robust, advanced, reliable, and simple with good maneuverability. Many researchers have been developed mobile robots to achieve these goals [1-3]. The applications are varied from rescuing to vacuums cleaning, from industrial to domestic. This project is to develop a mobile robot platform suitable for using in hospital as well as household for robot-assist evaluation, caring and rehabilitation services for the elderly and post-stroke patients. The maneuverability with good traction control of the robot will be based on two-wheeled is concentrated in this phase. Wheels on the robot are the most common form of maneuverability as the mechanism is straight forward with high dependability and still very popular for the new development of mobile robots such as the moon rover [4].

With the previous research on electrical vehicles, many sophisticated control algorithms were developed [5-10]. To extend the range of applications of the mobile robot, those advance control algorithms are needed to apply in the robotic field. Many traction control algorithms also have been developed in the past [5-6, 11]. This research explores the possibilities of the traction control algorithm.

To achieve the optimum travel speed with wheels on any plane, slip prevention must be realized. Without traction control algorithm, a wheel can be slip while driven with the torque which surpasses the friction force between wheels and floor. Moreover, retaining traction is important for driving in the desired direction otherwise the robot may turn unintentionally. Without additional orientation sensor, to maintain the driving direction, traction control is necessary. The traction control algorithm used in this research is developed based on the previous work, which used the model following control method (MFC) [11]. In this work, the simplification of the MFC is presented to show that dynamic parameters are negligible to some extends. The traction control is tested on the mobile robot and compared the result of the ones with position and velocity control.

In this paper, there are 6 main section organized as follows: The description of the mobile robot used in this research is stated in section 2, including the wheel configuration and robot specification. Section 3 states the baseline control scheme, a kind of classic position and velocity control schemes. Section 4 states the importance of the traction, previous traction controller, and the derivation of the simple discrete time traction control. The experiment condition is stated in section 5. In section 6, the result of the experiments is shown and discussed. Then the conclusion and future works are in section 7.

2. The two-wheeled mobile robot

The two-wheeled mobile robot as shown in Figure 1 is driven by two brushless motors and planetary gears are used to reduce the speed by the gear ratio of 1:6, which form a differentially steered drive system. The driven wheels are made of aluminum with rubber tire and the 4 caster wheels are used as the passive support wheels. The robot is powered by two 12V VRLA batteries in series to obtain 24 voltages. The driven wheels always contact on the floor by a suspension mechanism made by springs. Table 1 shows the specifications of the robot.

The wheels' motion will be controlled with a position and velocity controller and a traction controller. Trapezoidal velocity trajectory is used as a reference to attain the desired velocity and acceleration. To maintain the orientation of the robot, both wheels must maintain the same velocity at all time.

Table 1. The specifications of the two-wheeled mobile robot.

Specifications	
Stall torque	78 Nm
Max. continuous torque	5.784 Nm
Maximum speed	1.38 ms^{-1}
Maximum power	260 W
Wheels diameter	200 mm
Encoder	2500×4 pulse
Dimension	57×50×30 cm
Sampling rate	2000 Hz

(a) (b)

Figure 1. (a) The two-wheeled mobile robot, (b) the wheel

3. Position and velocity control scheme

To realize position and velocity tracking accuracy, the controller was designed without taking dynamics of the mobile robot into consideration. Each wheel is controlled separately. The position control is achieved by position feedback from the encoder on each wheel. The velocity feedback is calculated by the derivative of the wheel encoder position and the feedback signal is sent to achieve the velocity control. Each feedback signal has each own gain parameter multiplier, resulting in the whole feedback system work as a PD controller, which is ready for trajectory tracking control for both position and velocity as shown in Figure 2.

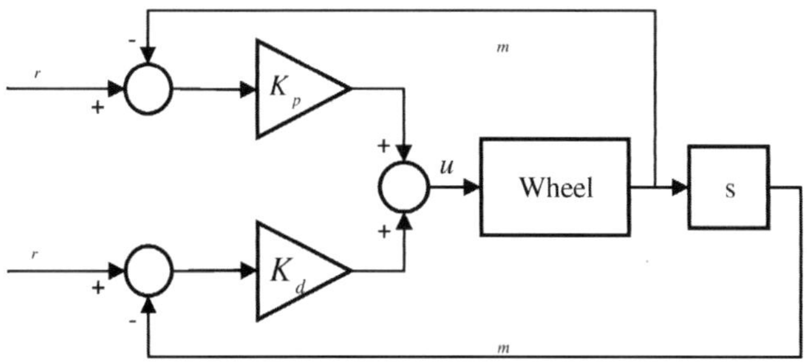

Figure 2. The block diagram of the position and velocity control scheme.

Reference position (θ_r) and velocity ($\dot{\theta}_r$) are determined by the trajectory planner then the position (θ_m) and velocity ($\dot{\theta}_m$) are fed back to perform closed-loop control with gain K_p and K_d. The wheel velocity calculated by performing derivation and low pass filtered on the position signal.

So, the control law is

$$u = K_p(\theta_r - \theta_m) + K_d(\dot{\theta}_r - \dot{\theta}_m) \quad (1)$$

4. Traction control scheme

Many works have been done in the traction control algorithm, e.g. traction control with slip estimator for a rocker-bogie robot [12], traction control with dynamic tire friction model [13], optimal slip ratio control and model following control [11].

The model following control (MFC) is simple to implement as it does not need the implementation of the tire-floor friction estimator or slip ratio estimator. MFC has also been verified through experiments for simple traction control [14].

To prevent the slip phenomenon, the modified implementation of the MFC is utilized in this two-wheeled robot. The block diagram of MFC is shown in Figure 3.

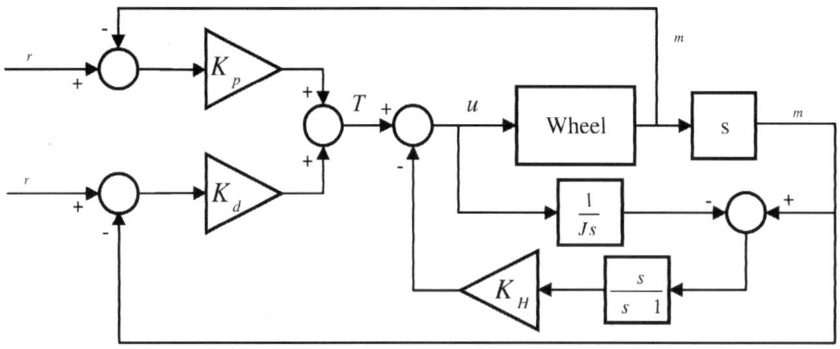

Figure 3. The block diagram of the model following control with position and velocity tracking control

The MFC prevent slip by reducing the output with the signal from a slip estimator. The output from the position and velocity controller (T) is used to estimate the wheel velocity and the slip is estimated by filtering the different of the measured velocity and the estimated velocity. The time constant of the high pass filter (τ), gain K_H, and system inertia (J) need to be chosen for this model. If slip, the sudden change of the measured velocity will be detected, and the slip will be reduced by the controller.

The output from the block diagram is

$$u=\left[T\left(s+\tau^{-1}\right)-K_H\dot{\theta}_m s\right]\left(s+\tau^{-1}-K_H J^{-1}\right)^{-1} \tag{2}$$

For implementation, the bilinear transformation was utilized. The bilinear transformation is a mathematical relationship which is used to convert the complex Laplace domain (s-domain) into z-domain. Using the substitution

$$s=\frac{2}{t}\frac{z-1}{z+1} \tag{3}$$

Where t is the system sampling time and z is the complex function variable.

And then

$$u=\frac{T\left[2(z-1)+\tau^{-1}t(z+1)\right]-K_H\dot{\theta}_m 2(z-1)}{\left[2(z-1)+\tau^{-1}t(z+1)-t(z+1)K_H J^{-1}\right]} \tag{4}$$

If the sampling time t is small enough, the part with t can be neglected

$$u=uz^{-1}+T\left(1-z^{-1}\right)-K_H\dot{\theta}_m\left(1-z^{-1}\right) \tag{5}$$

with

$$T=K_p(\theta_r-\theta_m)+K_d(\dot{\theta}_r-\dot{\theta}_m) \tag{6}$$

After the control law is simplified by removing the part with a very small coefficient, the high pass filter time constant and the system inertia are neglected, making the control law dynamically independent.

The output from the position and velocity controller T is used together with the wheel velocity signal $\dot{\theta}_m$, calculated from the derivative of θ_m, to make a slip prevention controller. The gain K_H is chosen to achieve the traction force.

5. Experiments

To verify the traction control law, both controllers are tested under the same experimental environment and parameters. The trajectory used in the experiment was generated using Trapezoidal velocity trajectory planner. The maximum velocity and acceleration are 1.3 ms^{-1} and 1.5 ms^{-2} respectively.

The mobile wheels are made of rubber and the floor is made of polished stone. The trajectory of each wheel of the robot is generated with the same parameter, that is, same position, same velocity, and same acceleration at any point in time. Ideally, the robot will move straight if slips did not occur.

6. Results and discussion

Figure 4 shows the result form the position and velocity controller and the traction controller. Figure 5 shows the timeline of the motion of the robot. The robot moves from the starting position at $t = 0\ sec$

IOP Conf. Series: Materials Science and Engineering **707** (2019) 012006 doi:10.1088/1757-899X/707/1/012006

to the position of the stop at $t = 3\ sec$ (right to left). The robot in Figure 5a was driven with the position and velocity controller and the same robot in Figure 5b was driven by the traction controller. It is assumed that if both wheels exert the same amount of traction force, the robot will move straight. But if either of wheels slips, the force from each tire will be unbalanced and the robot will turn unintentionally.

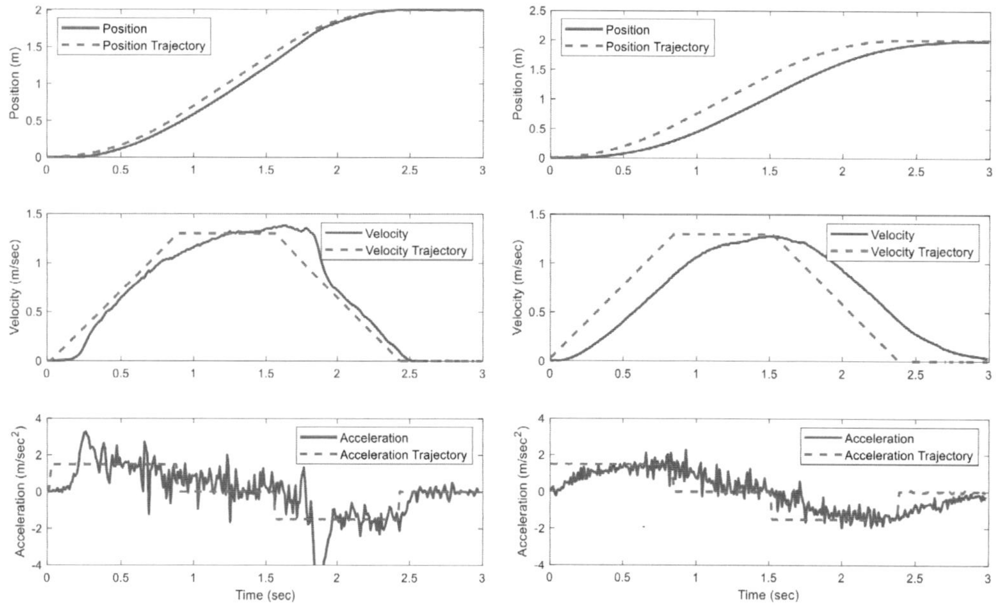

(a) (b)

Figure 4. The Trapezoidal velocity trajectory tracking by (a) the position and velocity controller, (b) the traction controller

(a) (b)

Figure 5. The motion of the robot while driven with (a) the position and velocity controller, (b) the traction controller

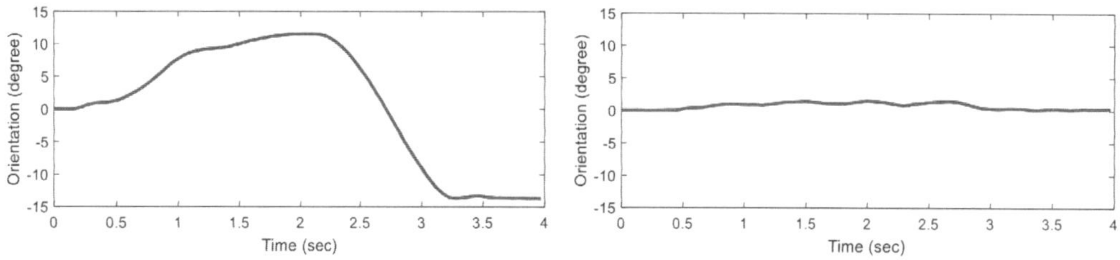

(a) (b)

Figure 6. The orientation of the robot with (a) the position and velocity controller, (b) the traction controller

In Figure 4a, the spike of the acceleration at the beginning shows that a wheel is slipped, corresponding with the robot in Figure 5a and Figure 6a. The slip causes the robot to turn and drive off the course. The slip also occurred at the end of the motion, generated by the deacceleration, which make the robot turn again but in the opposite direction.

The traction control law result was evaluated in Figure 4b. The velocity gradually increases without the spike in acceleration signal. The traction control was able to suppress the slip. From the robot motion in Figure 5b and the orientation in Figure 6b, the slip did not happen as the robot drive in a straight line and did not have any observable rotation motion.

The orientation of the robot shown in Figure 6 calculating from the yaw rate which measured with the 3DM-GX2 gyro sensor.

7. Conclusion and future work

The slip occurred in the first position and velocity controller experiment, noticeable from the spike of the acceleration at the beginning and the motion made by the robot and was managed to suppress by the modified MFC traction control law, which means the simplified MFC traction control works.

The MFC has proven effective on both EV and wheeled mobile robot but still required suitable dynamic parameters for implementation. Nevertheless, in the discrete domain, the MFC controller can be made simpler by the neglected some terms in the control law with sampling time multiplier. This approximation eliminates the dynamic parameter. The remaining terms in the control are related to state variables and gain parameter. With this simplification, the traction control is still successful according to the experimental results.

Future work will involve the estimation of the maximum traction force on any surface-pair using traction control on the wheeled mobile robot. Exploring other sophisticated traction controllers will be also considered.

8. References

[1] J. Borenstein, H.R. Everett, L. Feng, D. Wehe, *J. Robot. Syst.* **14(4)** 231–249 (1995).
[2] C. Canudas de Wit, *Control Probl. Robot. Autom.* 151–175 (2005).
[3] W. Poomarin, R. Chancharoen, V. Sangveraphunsiri, *Int. J. Mech. Eng. Robot. Res.* **5(1)** 11–16 (2016).
[4] http://www.cnsa.gov.cn/english/n6465652/n6465653/c6805105/content.html (2019).
[5] K. Fujii, H. Fujimoto, *IEEE PCC'07* 688–693 (2007).
[6] G. A. Magallan, C. H. De Angelo, G. O. Garcia, *IEEE Transactions on Vehicular Technology* **60(2)** 369–380 (2011).
[7] Y. Furukawa, N. Yuhara, S. Sano, H. Takeda, Y. Matsushita, *Vehicle system dynamics* **18(1-3)** 151–186 (1989).
[8] E. Ono, Y. Hattori, Y. Muragishi, K. Koibuchi, *Vehicle system dynamics* **44(2)** 139–151 (2006).
[9] S. Oh, K. Kong, Y. Hori, *Mechatronics* **24(8)** 1101–1111 (2014).
[10] K. Nam, S. Oh, H. Fujimoto, Y. Hori, *IEEE Transactions on Industrial Electronics* **60** 988–1000 (2013).
[11] Y. Hori, Y. Toyoda, Y. Tsuruoka, *IEEE Trans. Ind. Appl.* **34(5)** 1131–1138 (1998).
[12] M. Thianwiboon, V. Sangveraphunsiri, *Robocup 2005: Robot Soccer World Cup Ix* 682–690 (2006).
[13] C. Canudas de Wit, P. Tsiotras, *Proceedings of the 38th IEEE Conference on Decision and Control*, **4(Dec)** 3746–3751 (1999).
[14] J. H. Choi, S. Oh, *IEEE International Symposium on Industrial Electronics* **2018–June** 1075–1080 (2018).

IOP Conf. Series: Materials Science and Engineering **707** (2019) 012007 doi:10.1088/1757-899X/707/1/012007

Implementation of an automatic control system on a roaster of peruvian andes grains

Luis Garcia[1], Mario Lupaca[1], Diego Cardenas[1] and Karol Munoz[1]

[1] Mechatronic Engineering Specialty, Pontifical Catholic University of Peru, Lima, Peru

Abstract. This paper presents the results of the automation of a roaster of grains grown in the Peruvian Andes. The research work consisted in developing an automatic control system for the temperature and speed of rotation of the drum of a gas roaster machine with mechanical-electrical control. The new system is based on a PLC with a proportional control algorithm for temperature and an open loop for drum rotation, which allows the grain temperature to develop, in time, the predefined curves for its 3 types: corn, broad beans and soybeans. The cultivation of these in the various microclimates along the Andes Mountains gives them unique characteristics that condition the proper roasting to the operator's experience. To that end, a matrix was developed with parameters whose values were determined by tests conducted by expert operators in manual roasting. The experimental results showed that the roasted beans reached the desired quality characteristics such as humidity, color, aroma and flavor, as well as a decrease in the duration of the process and in fuel consumption, as well as the reduction of waste for over-roasting.

1. Introduction

The roasting process of Andean grains comprises a series of stages, each of which is defined by certain values of temperature, time and humidity level of the grain similar to the roasting process of coffee beans [1]. The automation of this process requires the control of certain variables, which depending on the characteristic of each grain will determine the adjustment of certain parameters for proper roasting. Because the response of the temperature of the grains to an increase in the reference signal is predominantly slow, it is feasible to use a proportional controller, which can reach the desired temperature in a suitable time, and maintain it with an error in state stable relatively small. The tests carried out with the objective of establishing the roasting parameters allowed to determine a roasting curve for each type of grain, being able to establish a thermal path line through which the grain passes during its roasting, being that between each reference signal adjusted by stage there is a transition stage that coincides with the curve that follows the grain temperature during the rise time, allowing a gradual increase in grain temperature. Following the roasting process stages on coffee beans tested on [1]; three stages were defined for the roasting of the Andean grains, after the preheating, a temperature uniformization stage is started for the grains recently admitted to the drum, in a second stage the temperature is increased to start the grain drying, and a final stage that seeks to active aromatic components in a phenomenon similar to the caramelization of sugars, which is characterized by a roasting of the external part of the grain in order to refine its aroma and color. The tests for the determination of the roasting curves for each grain allowed to establish empirically the proportionality constant of the control system implemented in a semi-industrial roaster of mechanical-electrical control, on which the various elements were adapted, such as: a control panel , a PLC with analog inputs and outputs, temperature sensors, gas pressure sensors, proportional valves for LPG, and a frequency converter, in addition to modifications in the mechanical and electrical elements in order to obtain a roaster whose automatic operation is programmable by the user. As a result, all these modifications have made it possible to standardize the roasting of the grains, reduce the processing times and the over-roasting that impaired the nutritional potential of the grain.

2. Automation of the grain roaster

In this chapter is explained the full operation of the roaster, the description of the temperature control system, the speed control system and the user interface for configuration and monitoring the process.

2.1. Operation of the roaster

The roaster shown in figure 1 consists of a horizontal cylindrical drum with capacity for 44 pounds of grains, two burners located under the drum that divides it into two zones of temperature control, an electric motor to rotate the drum on its own axis and a manual mechanism to tilt the drum after the grains are roasted. The roaster must be preheated before placing the grains in the drum. Then, the grains are added to the rotating drum and roasted. Finally, the roasted grains are removed from the drum due to the manual action of the mechanism that allows the drum to tilt. The roasting temperature, the roasting time and the centrifuge speed depend on the type of grain.

Figure 1. Main view of automatic roaster.

2.2. Temperature control system

In processes that involve temperature, it is important to use some type of control to reach a desired or defined temperature. For example, in the work carried out by Schiller, S [2], they compare the ON / OFF control and the proportional control to determine which is more efficient with the actuator being a pump, the plant is a solar collector and the sensor is a thermostat. They conclude that in most cases, the proportional control gives a greater efficiency to reach the desired temperature.

A more specific case, in which it is required to control the temperature in a meat oven, is found in the study conducted by Mielche, M [3]. In it, 4 different types of control are applied on the temperature variable with some modifications in the curves of the sensed value. The objective of its control system is to obtain the quality parameters without prolonging the heating time too long. The best type of control assigned is the so-called TDR, since it offers a smoother heat treatment in a shorter cooking time. TDR is based on the direct assignment of the supplied voltage in a short period of time and then a proportional control is applied.

Finally, in an experiment conducted by Chhaiya, Deep & Kathiriya, Rohit [4], a proportional temperature control was applied to a rotary roaster of food or products based on rice. After the implementation of its design, it is possible to study the accuracy of the type of control applied.

The roaster of the present study has two thermocouples in the zones of the drum to measure the temperature and contains a pair of solenoid valves to control the flow of gas in the burners. The sensors measure the temperature of the two zones in the drum in real time and send the information to the PLC. The scientific articles studied for temperature control and specifically those related to food treatment are taken into account. For this reason, a proportional control is applied.

IOP Conf. Series: Materials Science and Engineering **707** (2019) 012007 doi:10.1088/1757-899X/707/1/012007

2.3. Speed control system

It has been investigated how speed affects machines that use a rotating drum mechanism for the treatment of food. For example, the work presented by Kaensup, W [5] consists of a chili-drying machine that uses a speed variator in the rotating drum, they conclude that the variation is crucial to eliminate the effects in the dead zones, that is, in periods of time where the process starts or ends. In addition, the speed varies according to different types of foods that are used. The speed of rotation is controlled by an electronic speed variator.

Another work presented by M. Delele [6] indicates that the speed of rotation of the drum affects the operation of a rotary drum in two ways. On the one hand, it affects the production capacity, and, on the other hand, it is an important factor when determining the effect of the dead zone in a similar way to the article presented previously [5]. Additionally, in the study by A. Olasumboye [7] where a rotating drum mechanism is also used, it uses the method of varying the speed of the drum and uses an OMRON frequency variator.

Based on these mentioned investigations, the use of a frequency inverter was implemented to modify the speed of the spin drum according to the type and quantity of grain that is being roasted. The roaster has a gear motor of 1 HP and a speed of output of 60 rpm. In addition, it has a mechanical transmission by gears and chain. The pulleys have a number of teeth in a ratio of approximately 1: 2 because the speed delivered by the electric motor must be reduced and the toaster must rotate at a maximum speed of 29 rpm. A frequency variator of the SIEMENS brand is used to control the speed system, which is selected according to the power of the motor and the load it moves.

2.4. Interface for configuration and monitoring

The system controller is an S7-1200 PLC from the SIEMENS brand, it is used to control the temperature and speed variables by using sensors and actuators distributed along the roaster. The operating interface between the user and the system is a 7" touch screen from the SIEMENS brand. The communication with the controller is done through an RJ45 cable and programmed in the STEP 7 TIA PORTAL software. The interface is configured so that, when the process is started, the user can decide whether to perform an automatic or manual roasting as shown in figure 2.

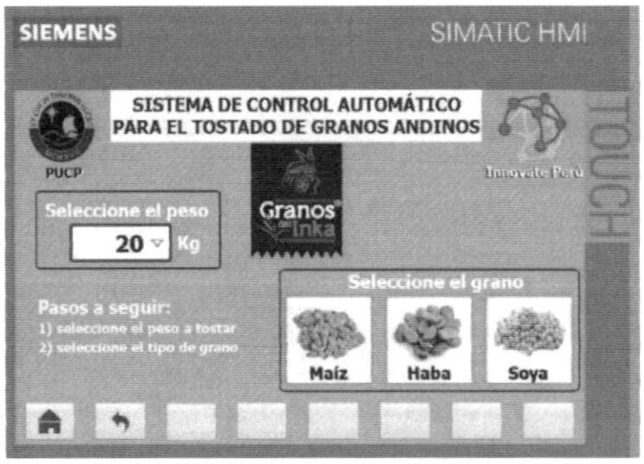

Figure 2. Image of the selection screen for weight and type of grain for automatic roasting.

3. The process of roasting Andean grains

In order to meet the desirable characteristics, a grain roasted for the production of pre-cooked or instant food requires certain physical and chemical modifications.

3.1. Chemical changes

There is a complex chemical process that triggers multiple effects on the chemical structure of a grain. This is known as the Maillard reaction [8].

3.1.1. Maillard reaction. The implemented toasting system maintains a temperature between 210ºC and 235ºC for a period of 12 minutes; these values depend on the type of grain. After that lapse of time, the Maillard reaction occurs, which can be perceived by the appearance of dark spots and cracks in the surface of the grain (see figure 3). This process is characteristic of toasting processes with high carbohydrate and protein content [8] and subjected to temperatures above 120ºC.

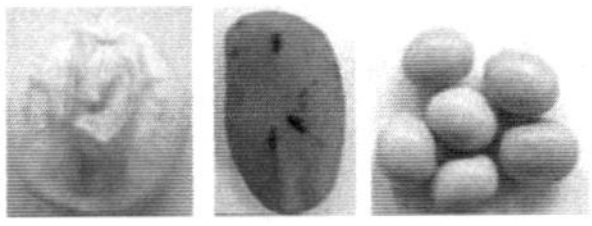

Figure 3. Maillard reaction in corn, broad bean, and soybeans respectively.

3.1.2. Activation of aromatic components. The chemical processes between the proteins and the sugars triggered at the beginning of the Maillard reaction release aromatic elements [9], which are accentuated in the last stage of roasting. In it, it continues with the reduction of moisture and water activity in the grain, which will allow the flours of these grains to be stable under environmental conditions.

3.2. Physical changes

The characteristics of the grains grown in the Andes are very diverse in terms of size, shape and level of humidity, because they are collected from different areas of cultivation, with diverse climatic conditions and sun dried by hand.

3.2.1. Drying process by moisture reduction. The heat applied on the grains releases water particles in the form of vapor. In turn, the decomposition and evaporation of certain organic acids cause the texture of the grain to change, towards a slightly rough surface and with cracks due to the rupture of the cell walls of the grain during the escape of the water molecules [8].

3.2.2. Modification of mass and density. Water vapor and carbon dioxide, as a consequence of the roasting process, generate high pressures inside the grains, which modifies their cellular structure, increasing the volume from the decrease in the thickness of the walls of the cells [8]. This is noticeable mainly in the grains of corn, a little less in the wheat and in the broad bean, with the consequent loss of mass by drying and reduction of the density after the increase in volume. In the measurements made, a reduction of the mass in the grain of the order of 7% is observed for the case of corn, with an average volume increase of 9%, while the broad beans lose 15% of mass and soybeans 3% after the roasting process.

3.2.3. Color variation. During the Maillard reaction, there is a change in color: the roasted grain acquires a brown color [10], although in the case of soybean there is not much difference. We can appreciate in figure 4, the change of color that the corn grains experience during toasting, with the presence of slightly darker areas due to the irregularity of the grain.

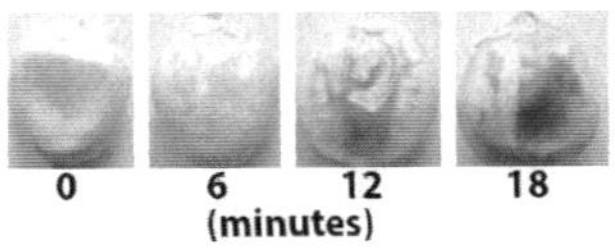

Figure 4. Darkening of corn color during automatic roasting.

Both the reduction of humidity and the low water activity present in the Andean grains after roasting facilitate the post-roasting processes, such as grinding.

4. Roasting matrix

It is known that, during roasting, the grains undergo various physical and chemical changes. The way in which these affect the food characteristics of the grain -ie nutritional value, flavour, aroma and color- will depend on the time of exposure to heat and the temperature at which they will be subjected. Considering these factors, a matrix was designed with parameters of speed, temperature and time, from which certain rules were established for the behaviour of the roaster in each of the three stages defined in the process of automatic roasting: a first stage for homogenize the temperature of the grains after loading, a second stage characterized by the loss of moisture, and the third stage where the Maillard reaction occurs.

5. Results

The automatic grain control system allowed the levels of humidity and acidity in the grains to be reduced after roasting. This last factor reduces the feeling of bitterness in food. The moisture reduction, the water activity (Aw) and the level of acidity are shown in table 1.

Table 1. Variation in percentage of the properties of the beans before and after roasting

	Moisture		Aw		Titratable acidity	
Grains	Before	After	Before	After	Before	After
Corn	15.32	8.55	0.65	0.38	0.24	0.13
Soybean	11.25	7.63	0.66	0.46	0.62	0.37
Broad bean	14.33	3.27	0.63	0.20	0.63	0.40

The new values of moisture and acidity were reached after roasting through the stages defined according to the values shown in table 2, where the system's operating configuration is shown.

Table 2. Configuration parameters of the roasting system

<table>
<tr><th colspan="2">Operational Parameters</th><th>Broad bean</th><th>Soybean</th><th>Corn</th></tr>
<tr><td colspan="2">Speed of rotation (RPM)</td><td>26</td><td>25</td><td>26</td></tr>
<tr><td colspan="2">Equipment load (Kg/batch)</td><td>20</td><td>20</td><td>20</td></tr>
<tr><td rowspan="3">Maximum Temperature (°C)</td><td>Stage 1</td><td>180</td><td>180</td><td>170</td></tr>
<tr><td>Stage 2</td><td>235</td><td>210</td><td>230</td></tr>
<tr><td>Stage 3</td><td>245</td><td>240</td><td>240</td></tr>
<tr><td colspan="2">Time (minutes)</td><td>29</td><td>15</td><td>18</td></tr>
</table>

Two zones were determined in the drum, in order to standardize the roasting of the grains, due to the disturbance that is generated in the door of grain entry. To compensate for this distortion, two loops of proportional control were implemented on the temperature of the roasting drum, one for each zone. In a stage prior to the entry of the grains, the preheating of the drum is performed whose maximum temperature will depend on the type of grain. As shown in figure 5, 6 and 7, the path of the temperature curves of the grains from the measurements of the sensors in zone 1 (light blue) and in zone 2 (orange).

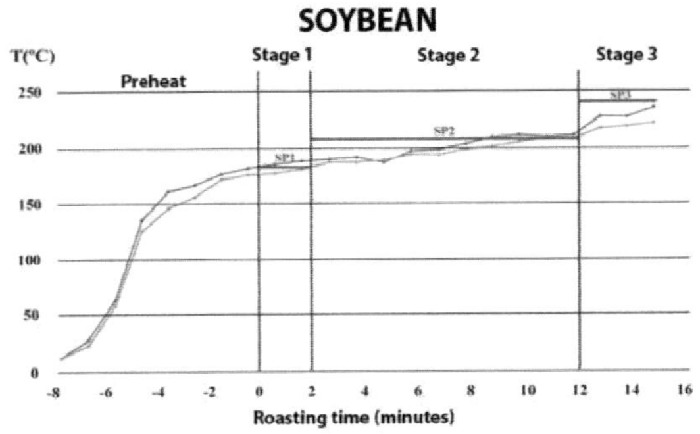

Figure 5. Curves of roasting of soybean.

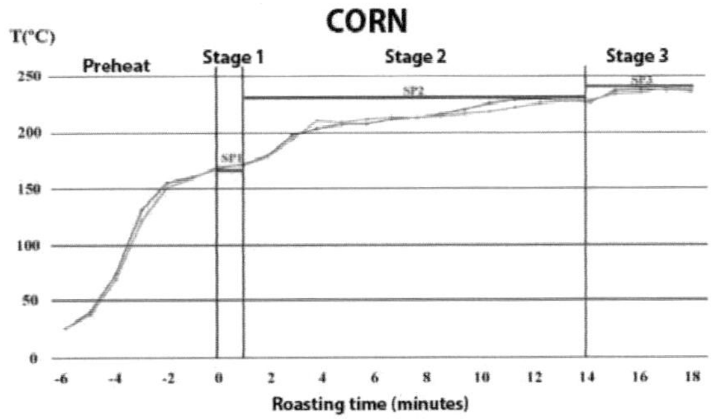

Figure 6. Curves of roasting of corn.

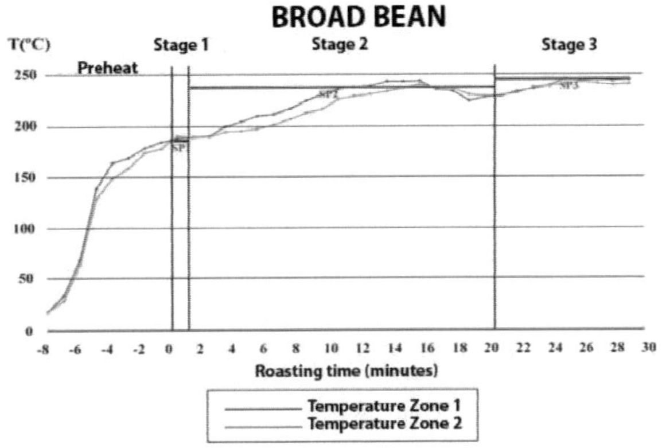

Figure 7. Curves of roasting of broad bean.

6. Conclusions

The selection of sensors and control elements was made based on the operating conditions of the roaster. In the particular case of temperature sensors, they must work non-invasively on the grains, to ensure safety and not generate disturbances on the process. The roasting of grains, after applying a temperature control, is shown uniform for the 44 lb. avoiding internal burn and better use of the material entered. The configuration parameters of the equipment for roasting broad beans, soybeans and corn allowed to obtain satisfactory and standardized results from the physical aspect (uniformity of color) and chemicals (% moisture, water activity and% acidity), getting grains or roasted seeds suitable for the following transformation processes. It was found that the temperature during the roasting of the grains and seeds had an error in stable state of less than 2%, that is to say in general rules it was kept within a range of ± 5 ° C in relation to the programmed temperature. Three stages of roasting were defined: a first short-term stage that allows the cold entering grains to acquire a uniform temperature; a second stage of greater duration, which favors the loss of moisture of the grain, as well as its internal cooking (the temperature curve must be gradually rising to avoid burning the external part of the grain); and finally a third stage in which the change of color and the aroma of the roasted grain are manifested. The development of this machine allows to improve the efficiency in the roasting process of the grains grown in the Andes. In Peru, this productive sector is mostly occupied by small companies that would be favored by increasing their competitiveness in the Andean grain derivatives market.

7. References

[1] C. Wrigly, H. Corke, K. Seetharaman and J. Faubion. "Encyclopedia of FOOD GRAINS". 2016
[2] Wen-Keng Tseng, King-Chu Hung, Frank Yih-Fong Tzeng,and Tsung-Ching Wu. "Taguchi Analysis for Quality Stabilization of Coffee Roast".2017

[3] S. Schiller, M. Warren and D. Auslander, "Comparison of Proportional and On/Off Solar Collector Loop Control Strategies Using a Dynamic Collector Model." 1980.
[4] M. Melche, "Effect of different temperature control methods on heating time and certain quality parameters of oven-roasted beef" 1992.
[5] D. Chhaiya and R. Kathiriya, "Proportional Temperature Control for food grade rotary roaster furnace using RF transceiver. International journal of research in engineering & advanced technology" 2016.
[6] W. Kaensup, S. Chutima and S. Wongwises, "Experimental study on drying of chilli in a combined microwave-vacuum-rotary drum dryer" 2002.
[7] M. Delele, F. Weigler and J. Mellmann, "Advances in the Application of Rotary Dryer for Drying of Agricultural Products: A Review" 2014.
[8] A. Olasumboye, O. Koya, "A Rotary Separator for the Dry Mixture of Palm Kernel and Shell" 2014.
[9] A. Jaiswal, "Food processing technologies: Impact on Product Attributes" 2017.
[10] H. Nursten, "The Maillard Reaction: Chemistry, Biochemistry and Implications" 2005.

Motion control of self-driving vehicles on rough roads

I Shapovalov[1,2], S Gafurov[1] and A Klimchik[1]

[1]Center for technologies in Robotics and Mechatronics components, Innopolis University, 1 Universitetskaya Str., Innopolis, Russia
[2]Author to whom any correspondence should be addressed

Abstract. Control of self-driving vehicles is one of the-up-to-date tasks. Vehicles are non-linear dynamic systems with a large number of parameters. When it is supposed to move on a rough road with a lot of abrupt turns, significant unpredictable and unmeasurable disturbances emerge and the control quality degrades significantly. We implemented model predictive path integral (MPPI) control to deal with significant unmeasured disturbances. To approve efficiency of the proposed controller, we conducted simulation of complex car motion on a plane under the control of PID, and MPPI controllers. Performance of controllers is comparatively equal in stable modes, when reference position and velocity do not change. MPPI controllers outperform PID ones in unstable modes due to their ability to predict a vehicle behavior. Also, MPPI is the best choice for the rejection of significant external disturbances.

1. Introduction

Self-driving cars navigation is provided by a number of subsystems such as localization, routing and path-planning subsystems as well as a control subsystem and a perception [1]. Autonomous car control, route generation and path-planning are provided by a number of methods and algorithms. And usually, these tasks are considered separately. Moreover, oftentimes path-planning algorithms do not take into account a car dynamics model as there is a number of scenarios where it can be left out of the account. In such scenarios, a car kinematics model is only used for path-planning and grid/graph search techniques are used for global routing [2], [3]. However off-road driving scenarios as well as driving in an urban environment with a number of maneuvers require a highly-detailed dynamic model of the car in order to perform low-level path-planning and to ensure its feasibility in real-time.

Methods based on Model Predictive Control (MPC) and PID control are widely used for self-driving cars. MPC is a model-based approach which has shown good results for car navigation in real-time [4], [5]. MPC allows adjusting vehicle parameters according to driving conditions. It requires knowing of several operational modes and corresponding internal models of a vehicle to shift between models and adjust parameters of the models if it is necessary [6], [7]. Linearization of a non-linear model was performed in research [8] in order to control a vehicle that moves at a constant speed. The control is based on the nested structure of two PID regulators. The obtained results showed that the proposed approach is better than the standard MPC method [9].

However self-driving vehicles require a new type of control - integral control of its various subsystems in a case when vehicle dynamic properties are crucial to be considered. The Model Predictive Path Integral control method has been shown to have more potential for self-driving car implementation [10]. It allows considering the path-planning and control subsystems as a whole. MPPI controllers require a raster costmap for environment representation. The research [11] has investigated the efficiency of the MPPI algorithm for relatively simple driving conditions. Our

IOP Conf. Series: Materials Science and Engineering **707** (2019) 012008 doi:10.1088/1757-899X/707/1/012008

research focuses on the investigation of the MPPI method efficiency and its working performances compared to standard control methods.

2. Problem statement

We consider an electric car Kia Soul EV as a control object in this paper. The problem statement is to move autonomously on a rough surface track with several turns and short straight segments similar to country roads. At the same time, we suppose the road surface inclination is always zero. The motion speed varies in a range from 25 to 40 km/h. Such requirements to motion predetermine strong influence of various external disturbances that can be represented as a combination of step, sine and Gaussian-noise signals.

Planning the route of a vehicle is another complex control problem that deserves close attention and efforts that are outside the scope of the paper. We suppose that some planning subsystem forms the reference speed V_{ref}, lateral position y_{ref} and car orientation ψ_{ref} in a road coordinate system for the motion controller. The planner solves all problems related to providing trajectories for obstacles and potholes avoidance.

To describe motion of the self-driven car we use two coordinate frames as it is shown in figure 1. The first frame is vehicle-fixed and the second one is an earth-fixed coordinate frame. The origin of the vehicle-fixed frame is in the car centre of mass.

3. Model of vehicle

Most of research papers about self-driving vehicles discuss a single-track dynamics model [12] as it allows describing basic car dynamics without complex expressions with a large number of parameters. On the other hand, dual-track models allow considering a number of non-linearities occurring in extreme or close-to-extreme operational modes. The second type of models includes a number of parameters and their integration requires a great amount of calculation. That makes it inexpedient to implement such models in onboard computers to achieve control goals. But simulation of dual-track models on stationary computers provides an estimation of motion control and signal processing systems quality with the accuracy necessary for real-world problems.

Let us consider a vehicle with two axles for the description of its longitudinal and lateral motion with orientation variation. The schematic representation of the vehicle with used coordinate frames is given in figure 1.

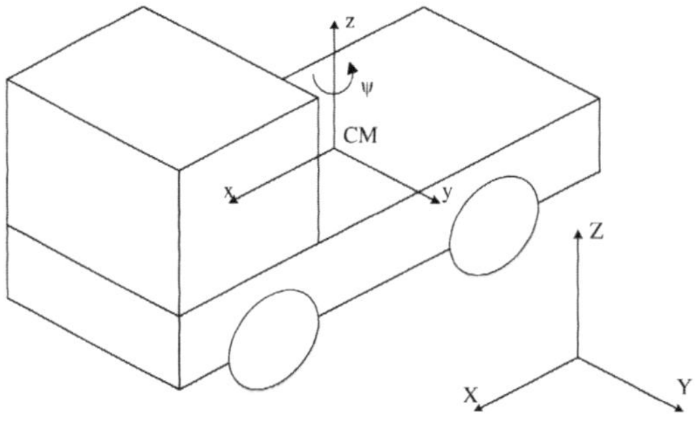

Figure 1. Coordinate frames used for vehicle description.

Dynamics of a vehicle as a solid body can be represented in the form of the following equations [13]:

$$\begin{gathered}\ddot{x}=\dot{y}r+\frac{F_{\mathrm{xfl}}+F_{\mathrm{xfr}}+F_{\mathrm{xrl}}+F_{\mathrm{xrr}}+F_{\mathrm{xext}}}{m}\\ \ddot{y}=-\dot{x}r+\frac{F_{\mathrm{yfl}}+F_{\mathrm{yfr}}+F_{\mathrm{yrl}}+F_{\mathrm{yrr}}+F_{\mathrm{yext}}}{m}\\ \dot{r}=\frac{a\left(F_{\mathrm{yfl}}+F_{\mathrm{yfr}}\right)-b\left(F_{\mathrm{yrl}}+F_{\mathrm{yrr}}\right)+\frac{w_{\mathrm{f}}}{2(F_{\mathrm{xfl}}-F_{\mathrm{xfr}})}+\frac{w_{\mathrm{r}}}{2(F_{\mathrm{xrl}}-F_{\mathrm{xrr}})}+M_{\mathrm{d\,z}}}{I_{zz}}\end{gathered} \tag{1}$$

IOP Conf. Series: Materials Science and Engineering **707** (2019) 012008 doi:10.1088/1757-899X/707/1/012008

where m is the vehicle mass; a, b are the distances to the front and rear axle correspondingly from the point of CM projection on the axle plane; w_f, w_r are the front and rear track widths; Izz is the vehicle body moment of inertia about the vehicle-fixed z-axis.

Forces applied to the attachment points of wheels are defined as:

$$F_{\mathrm{xflt}}=F_{\mathrm{xflinput}},\ F_{xfrt}=F_{xlrinput},\ F_{\mathrm{xrlt}}=F_{\mathrm{xrlinput}},\ F_{\mathrm{xrrt}}=F_{\mathrm{xrrinput}}$$

$$F_{\mathrm{yflt}}=-C_{\mathrm{yfl}}\alpha_{\mathrm{fl}}\mu_{\mathrm{fl}}\frac{F_{\mathrm{zfl}}}{2F_{\mathrm{znom}}},F_{\mathrm{yfrt}}=-C_{\mathrm{yfr}}\alpha_{\mathrm{fr}}\mu_{\mathrm{fr}}\frac{F_{\mathrm{zfr}}}{2F_{\mathrm{znom}}},F_{\mathrm{yrlt}}=-C_{\mathrm{yrl}}\alpha_{\mathrm{rl}}\mu_{\mathrm{rl}}\frac{F_{\mathrm{zrl}}}{2F_{\mathrm{znom}}},F_{\mathrm{yrrt}}=-C_{\mathrm{yrr}}\alpha_{\mathrm{rr}}\mu_{\mathrm{rr}}\frac{F_{\mathrm{zrr}}}{2F_{\mathrm{znom}}} \quad (2)$$

where F_{input} are the tractive forces emerging as a result of contact between the wheels and earth surface; F_{znom} is the nominal normal force applied to the vehicle axles used to estimate friction parameters during weight and load shift; $C_{\mathrm{y}\ i,j}$ is the cornering stiffness of wheels, μ is the wheel friction coefficient.

Slip angles α are defined based on the ratio of velocities in the vehicle-fixed coordinate frame and the front wheels steering angle:

$$\alpha_{\mathrm{fl}}=\mathrm{atan}\left(\frac{\dot{y}+ar}{\dot{x}+r\frac{w_{\mathrm{f}}}{2}}\right)-\delta_{\mathrm{f}},\ \alpha_{\mathrm{fr}}=\mathrm{atan}\left(\frac{\dot{y}+ar}{\dot{x}-r\frac{w_{\mathrm{f}}}{2}}\right)-\delta_{\mathrm{f}},\ \alpha_{\mathrm{rl}}=\mathrm{atan}\left(\frac{\dot{y}-ar}{\dot{x}+r\frac{w_{\mathrm{r}}}{2}}\right),\ \alpha_{\mathrm{rr}}=\mathrm{atan}\left(\frac{\dot{y}-ar}{\dot{x}-r\frac{w_{\mathrm{r}}}{2}}\right) \quad (3)$$

where δ_{f} is the steering angle of the front wheels.

The forces from equation (2) are included into equation (1) after their representation in the vehicle-fixed coordinate frame:

$$F_{\mathrm{xf}}=F_{\mathrm{xft}}\cos(\delta_{\mathrm{f}})-F_{\mathrm{yft}}\sin(\delta_{\mathrm{f}}),\ F_{\mathrm{yf}}=-F_{\mathrm{xft}}\sin(\delta_{\mathrm{f}})+F_{\mathrm{yft}}\cos(\delta_{\mathrm{f}}),F_{\mathrm{xr}}=F_{\mathrm{xrt}},F_{\mathrm{yf}}=F_{\mathrm{yft}} \quad (4)$$

4. Motion controller

Further, we give basic expressions needed to calculate MPPI control signals. To achieve better understanding of their inference, a reader can acquaint himself with [10].

Let's consider nonlinear system in the form

$$\mathrm{d}\mathbf{x}=\mathbf{f}(\mathbf{x}_t,t)\mathrm{d}t+\mathbf{G}(\mathbf{x}_t,t)\mathbf{u}(\mathbf{x}_t,t)\mathrm{d}t+\mathbf{B}(\mathbf{x}_t,t)\mathrm{d}w \quad (5)$$

The initial state vector of the system (5) is denoted as x0, and the terminal state at time T is $_{\mathrm{xT}}$. It should be noted that the system is affine in control and Brownian disturbance. We assume, that **G** and **B** are partitioned in such a way that we can consider actuated and non-actuated dynamics of the system. Also, there should be no correlation between noise affecting actuated and non-actuated parts of the system (5) with a stationary noise in the actuated part:

$$\mathbf{G}=\begin{pmatrix}0\\ \mathbf{G}_c\end{pmatrix},\ \mathbf{B}(\mathbf{x}_t)=\begin{pmatrix}\mathbf{B}_\alpha(\mathbf{x}_t) & 0\\ 0 & \mathbf{B}_c\end{pmatrix}$$

The disturbance can be represented as

$$\mathrm{d}w=\epsilon\sqrt{\mathrm{d}t}$$

In discrete form it looks like

$$\mathrm{d}\mathbf{x}_{tj}=(\mathbf{f}(\mathbf{x}_{tj},t_j)+\mathbf{G}(\mathbf{x}_{tj},t_j)\mathbf{u}_j)\Delta t+\mathbf{B}(\mathbf{x}_{tj},t_j)\epsilon_j\sqrt{\Delta t} \quad (6)$$

The cost function for estimation of control quality over the system trajectory is

$$\bar{S}(\tau)=\varphi(\mathbf{x}_{\mathrm{T}},\mathrm{T})+\textstyle\sum_{j=0}^{\mathrm{N}}\tilde{q}(\mathbf{x}_t,\mathbf{u}_t,\epsilon_t,t)\Delta t \quad (7)$$

where $\varphi(\mathbf{x}_{\mathrm{T}},T)$ is the terminal cost.

The second term in the right part of equation (7) is the sum of running costs consisting of control, state-dependent and stochastic components. We calculate the running costs according to the following expression

$$\tilde{q}(\mathbf{x}_t,\mathbf{u}_t,\epsilon_t,t)=q(\mathbf{x}_t,t)+\frac{1}{2}\mathbf{u}_t^{\mathrm{T}}\mathbf{R}\mathbf{u}_t+\lambda\mathbf{u}^T\mathbf{G}\frac{\epsilon}{\sqrt{\Delta t}}+\frac{1}{2}\lambda(1-\nu^{-1})\frac{\epsilon^T}{\sqrt{\Delta t}}\mathbf{B}_c^T(\mathbf{B}_c\mathbf{B}_c^{\mathrm{T}})^{-1}\mathbf{B}_c\frac{\epsilon}{\sqrt{\Delta t}} \quad (8)$$

where $\mathbf{R}=\mathbf{R}(\mathbf{x},t)$ is a positive definite matrix, λ is a constant establishing the relation between noise and the cost function S.

We take the state-dependent part of the running cost as

$$q(\mathbf{x},t)=\mathrm{k}(\mathbf{x}-\mathbf{x}_{\mathrm{ref}})^2 \quad (9)$$

The iterative form of the control signal $\mathbf{u}_j^*$ is as follows

$$\mathbf{u}_j^* = \mathbf{u}_j + \mathcal{H}^{-1}\mathcal{G}\left(E_{q_u^v}\left[\frac{\exp\left(-\frac{1}{\lambda}\tilde{S}(\tau)\right)\frac{\epsilon_j}{\sqrt{\Delta t}}}{E_{q_u^v}\left[\exp\left(-\frac{1}{\lambda}\tilde{S}(\tau)\right)\right]}\right]\right) \tag{10}$$

The terms

$$\mathcal{H} = \mathbf{G}_c^T(\mathbf{B}_c\mathbf{B}_c^T)^{-1}\mathbf{G}_c \qquad \mathcal{G} = \mathbf{G}_c^T(\mathbf{B}_c\mathbf{B}_c^T)^{-1}\mathbf{B}_c$$

If we approximate the expectation $E_{q_u^v}$ over a stochastic process q with a non-zero control input $\mathbf{u}$ and a changed variance $\mathbf{v}$ by a weighted average over sampled trajectories τ_i and take the dependency $\mathcal{H}^{-1}\mathcal{G} = \gamma$, the control (10) has the form

$$\mathbf{u}(\mathbf{x}_{t_i}, t_i)^* \approx \mathbf{u}\left(\mathbf{x}_{t_i}, t_i\right) + \frac{\sum_{k=1}^{K}\exp\left(-\frac{1}{\lambda}\tilde{S}(\tau_{i,k})\right)\delta\mathbf{u}_{i,k}}{\sum_{k=1}^{K}\exp\left(-\frac{1}{\lambda}\tilde{S}(\tau_{i,k})\right)} \tag{11}$$

The algorithm of MPPI control describing a consequence of calculations for obtaining the controller output signal can be found in [8].

5. Simulation

Efficiency of presented approach was checked in MatLab environment. As the parameters of the model (1) we used the exact or approximate parameters of Kia Soul EV given in table 1.

Table 1. Parameters of Kia Soul EV dynamics.

Parameter	Value
Mass m, kg	1513
Moment of inertia I_{zz}, kg·m^2	2066
Distance to front axle a, m	1.2
Distance to rear axle b, m	1.37
Height of center of mass h, m	0.15
Nominal normal force F_{znom}, N	7500
Front wheel cornering stiffness C_{yf}, N/rad	15000
Rear wheel cornering stiffness C_{yr}, N/rad	35000

The tire friction coefficient μ has the value of 0.72. We also add several restrictions on the vehicle dynamics. The steering angle rate of change is 6 degrees per sec with a corresponding sign. The speed rate of change is 10 m/sec2 with a sign corresponding to acceleration or deceleration mode.

We compared performance of the PID and MPPI controllers in various scenarios to reveal their advantages also with restrictions. For the MPPI controller we used the prediction horizon of 1 sec with 50 sample trajectories.

We presented results of using the PID and MPPI controllers during motion without external disturbances and rare changes in reference signals in figure 2. Figure a demonstrates changing of the vehicle longitudinal speed, and figure 2b shows the trajectories obtained by two compared control systems. As it can be seen the performance of controllers is similar. MPPI is faster a little, but by the price of overshooting in the lateral coordinate.

IOP Conf. Series: Materials Science and Engineering **707** (2019) 012008 doi:10.1088/1757-899X/707/1/012008

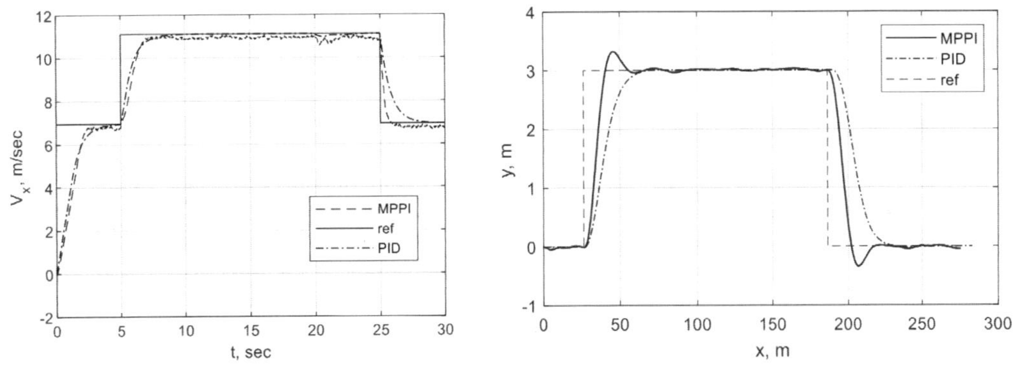

(a) (b)

Figure 2. Comparison of PID and MPPI controllers in mode without disturbances.

Next, we tried to model the conditions in which the vehicle moves on the rough country road. For that we included external disturbances in the vehicle model in the form of white noise of significant amplitude comparing with the vehicle throttle. Also, we used a sine form of the reference signal for the lateral vehicle position. Obtained results are presented on figure 3, figure 4 and figure 5. Special attention should be given to the results in figure 5 where we presented unfiltered throttle and steering control of MPPI in presence of disturbance. It is obvious that without post processing of such signal control could damage powertrain of the vehicle.

Trajectories in figure 4 demonstrate efficiency of MPPI control in transition modes as it can follow reference appropriately with small delay.

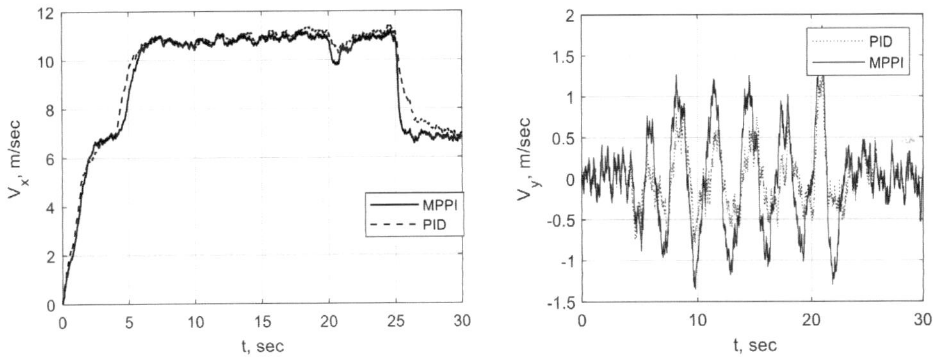

Figure 3. Longitudinal and lateral speed of vehicle in presence of disturbance.

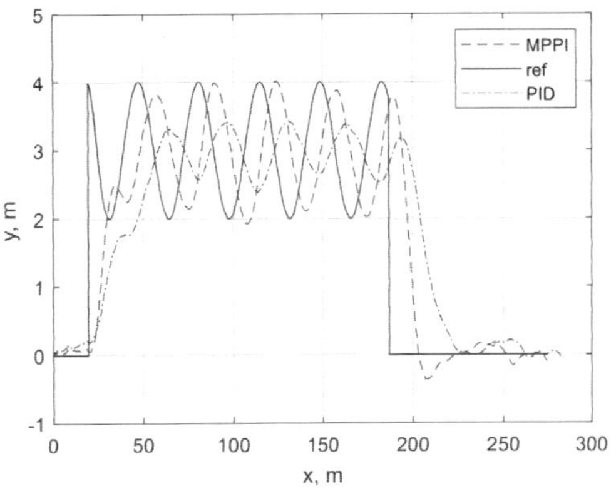

Figure 4. Trajectories of vehicle in presence of disturbance.

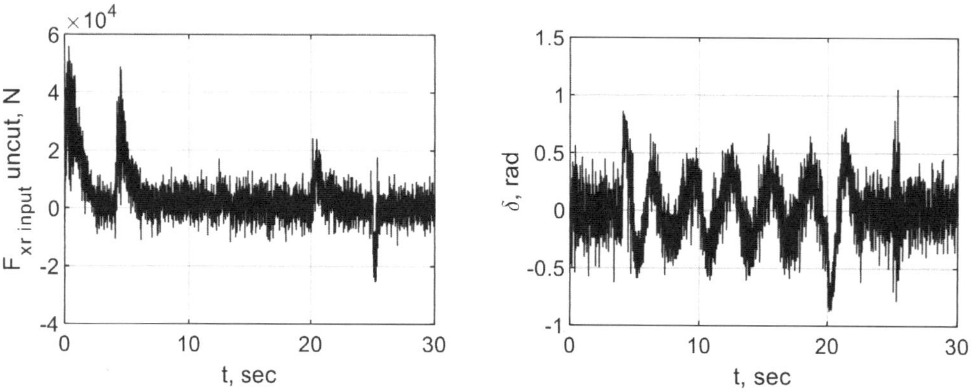

Figure 5. Unfiltered throttle and steering control of MPPI in presence of disturbance.

6. Conclusions and future work

Separate considering of self-driving car subsystems may lead to physically invalid routes as actual dynamics characteristics of the car are not taken into account. Developed in this research, the control method based on MPPI provides better behaviour of the vehicle. However considering of path-planning and control subsystems as a whole arises the problem of computational complexity and time delays that will be take into account in our further work.

7. Acknowledgments

This research was supported by the Russian Education Ministry under the Agreement No. 075-10-2018-017 on granting a subsidy from 11.26.2018 "Development of a commercial urban transport with the intellectual driver assistance system "City Pilot". The customer is a federal target program. The unique identifier of the project RFMEFI60918X0005. Industrial partner – KAMAZ.

8. References

[1] Chaaban J 2014 A Methodology for designing robust central/self-organising multi-agent systems *Int. J. of Computer Information Systems and Industrial Management Applications* **6** 571-81.

[2] LaValle S M and Branicky M S 2004 On the relationship between classical grid search and probabilistic roadmaps *Algorithmic Foundations of Robotics V. Springer Tracts in Advanced Robotics* vol 7, ed. Boissonnat JD., Burdick J., Goldberg K., Hutchinson S. (Berlin, Heidelberg: Springer).

[3] Latombe J C 1991 *Robot Motion Planning* (Boston, MA: Kluwer Academic Publishers).

[4] Khan M, Tahiyat M, Imtiaz S, Shoukat Choudhury M A A and Khan F 2017 Experimental evaluation of control performance of MPC as a regulatory controller *ISA Transactions* **70** 512-20.

[5] Worthmann K, Mehrez M W, Mann G K I, Gosine R G and Pannek J 2017 Interaction of open and closed loop control in MPC *Automatica* **82** 243-50.

[6] Pacejka H 2004 *Tire and Vehicle Dynamics* (Oxford: Butteworth Heinemann).

[7] Choi M, Oh J J and Choi S B 2013 Linearized Recursive Least Squares Methods for Real-Time Identification of Tire–Road Friction Coefficient *IEEE TRANS. ON VEHICULAR TECHNOLOGY* 7 **62** 2906-18.

[8] Marino R, Scalzi S and Netto M 2011 Nested PID steering control for lane keeping in autonomous vehicles *Control Engineering Practice* **19** 1459-7.

[9] MacAdam C 1981 Application of an optimal preview control for simulation of closed-loop automobile driving *IEEE Transactions on Systems, Man, and Cybernetics* **11** 393-9.

[10] Williams G, Drews P, Goldfain B, Rehg J M and Theodorou E A 2016 Aggressive Driving with Model Predictive Path Integral Control // 2016 IEEE International Conference on Robotics and Automation (ICRA), pp. 1433 – 40.

IOP Conf. Series: Materials Science and Engineering **707** (2019) 012008 doi:10.1088/1757-899X/707/1/012008

[11] Buyval A, Gabdullin A and Klimchik A 2018 Model Predictive Path Integral Control for Car Driving with Dynamic Cost Map. *Proc. 15th International Conference on Informatics in Control, Automation and Robotics* (*Porto*) vol 1 pp 248-54.
[12] Ackermann J 2002 *Robust Control* (London: Springer).
[13] Gillespie T 1992 *Fundamentals of Vehicle Dynamics* (Warrendale, PA: SAE, Inc.).

IOP Conf. Series: Materials Science and Engineering **707** (2019) 012009 doi:10.1088/1757-899X/707/1/012009

Parametric synthesis of a robust PID-controller for interval control system with aperiodic transient process

T A Ezangina[1], S A Gayvoronskiy[1] and I V Khozhaev[1]

[1] School of Computer Science and Robotics, National Research Tomsk Polytechnic University, Tomsk, 634034, Russia

Abstract. The paper is dedicated to a problem of synthesizing a PID-controller, which provides and aperiodical transient process of desired duration in a control system of the third order with interval parametric uncertainty. To solve the problem, a new method of synthesis was developed. It is based on a decomposing a characteristic polynomial of a considered system in two multipliers: dominant polynomial, which determines allocation of a dominant pole and, consequently, quality of a transient process in a considered system, and non-dominant polynomial, which determines allocation of all other system poles. In order to provide a desired allocation of system poles, parameters of a PID-controller are divided to unrestricted ones and dependent ones. The research resulted in algebraic expressions for calculating dependent parameters of a PID-controller and a system of inequalities for finding its unrestricted parameter. Developed method of a PID-controller parametric synthesis allows placing an interval real dominant pole according to desired control quality and placing all other poles of a system in a desired part of a complex plane to provide necessary degree of a robust pole domination.

1. Introduction

According to system pole domination principle, a control quality of a system is determined by a system pole, which is the nearest to an imaginary axis of a complex plane. Considering this, transient process aperiodicity can be providing by placing a real pole near an imaginary axis and placing all other poles far enough from the real pole to reduce their influence on a transient process quality.

Due to interval parametric uncertainty of a system, each pole of the system will not be constant and will migrate inside its allocation area on a complex plane. Dominant real pole will migrate along a real axis of a complex plane within the range $[\alpha_1,\alpha_2]$. Let us introduce a term robust domination degree. Robust pole domination degree is equal to distance between left limit of a domination pole and right limit of a non-dominant poles allocation area. Let us designate a robust pole domination degree as follows: $\gamma=[\eta,\alpha_2]$. Value α_1 determines a robust stability degree – a minimal stability degree of an interval control system (ICS).

Parametric synthesis of controllers, which provide a desired allocation of system poles according to poles domination principle, is a highly relevant problem. Papers [1] – [14] are dedicated to developing its solution. One of common approaches to simplifying a controller synthesis procedure is to use low-degree transfer functions of a control object instead of complex full-degree ones. The degree of transfer function can be reduced by ignoring insignificant parameters of a control object or a whole system.

The paper is dedicated to solving a problem of parametric synthesis of a PID-controller, which provides an aperiodic transient process of desired duration, for an ICS of the third degree. With the help of robust control theory and poles domination principle, the problem can be reduced to developing a synthesis method, which will allow calculating such values of controller parameters k_i, $i = 0,1,2$, which provide a dominant pole allocation in a desired range of a real axis of a complex plane and provide non-dominant poles allocation in a complex plane area Γ with a desired robust pole domination degree γ.

2. Characteristic polynomial decomposition

Let us assume that interval characteristic polynomial (ICP) of a control system is written as follows:

$$A(s) = \sum_{i=0}^{n} [a_i]s^i, \underline{a_i} < a_i < \overline{a_i}, \tag{1}$$

where n – is a degree of an ICP.

Let us write a dominant polynomial, which is determined by a dominant pole value:

$$Q(s) = s + [\alpha], \alpha > 0,\ \alpha_1 < \alpha < \alpha_2. \tag{2}$$

where α – is a real interval dominant pole.

ICP (1) division by dominant polynomial (2) results in a non-dominant polynomial $P(s)$ and a remainder R. Expression for non-dominant polynomial $P(s)$ of an ICP of third order can be written as follows:

$$P(s) = \sum_{i=0}^{2} [p_i]s^i = [p_2]s^2 + [p_1]s + [p_0],\ [p_i] > 0,\ \underline{p_i} < p_i < \overline{p_i}. \tag{3}$$

Coefficients of $P(s)$ can be calculated with the help of following expression:

$$\sum_{j=i+1}^{n} [a_j][\alpha]^{j-i-1}(-1)^{j-i-1} = [p_i]. \tag{4}$$

To calculate $\underline{p_i}$ coefficients, $\underline{a_j \alpha^{j-i-1}}$ must be substituted to (4), if $(-1)^{j-i-1} = 1$, otherwise, if $(-1)^{j-i-1} = -1$, $\overline{a_j \alpha^{j-i-1}}$ should be substituted. To calculate, following substitutions must be performed: $\underline{a_j \alpha^{j-i-1}}$, if $(-1)^{j-i-1} = -1$; and $\overline{a_j \alpha^{j-i-1}}$, if $(-1)^{j-i-1} = 1$.

Expression for calculating a R remainder can be written as follows:

$$R = \sum_{i=0}^{n} [a_i][\alpha]^i(-1)^i. \tag{5}$$

Let us assume, that a control object of an ICS is described by the transfer function $W_{oc}(s) = \dfrac{k}{[c_2]s^2 + [c_1]s + [c_0]}$, $\underline{c} < c < \overline{c}$,; transfer function of the PID-controller is written as follows $W_c(s) = \dfrac{k_2 s^2 + k_1 s + k_0}{s}$. Considering this, an ICP of the system can be written as follows:

$$A(s, \vec{k}) = [a_3]s^3 + [a_2(k_2)] \cdot s^2 + [a_1(k_1)] \cdot s + [a_0(k_0)], \tag{6}$$

where $[a_3] = [c_2]$, $[a_2(k_2)] = [c_1] + k \cdot k_2$, $[a_1(k_1)] = [c_0] + k \cdot k_1$, $[a_0(k_0)] = k \cdot k_0$.

On the base of (4), let us derive expressions for coefficients of a non-dominant polynomial, using designations from the ICP (6) (see Table 1)

Table 1. Coefficients of interval non-dominant polynomial $P(s)$ for ICP (6)

$[p_i]$	Coefficients of $P(s)$ polynomial
$\underline{p_2}$	$\underline{a_3} = \underline{c_2}$
$\overline{p_2}$	$\overline{a_3} = \overline{c_2}$
$\underline{p_1}$	$\underline{a_2(k_2)} - \alpha_2 \overline{a_3} = (\underline{c_1} + kk_2) - \alpha_2 \overline{c_2}$
$\overline{p_1}$	$\overline{a_2(k_2)} - \alpha_1 \underline{a_3} = (\overline{c_1} + kk_2) - \alpha_1 \underline{c_2}$
$\underline{p_0}$	$\underline{a_1(k_1)} - \alpha_2(\overline{a_2(k_2)} - \alpha_1 \underline{a_3}) = (\underline{c_0} + kk_1) - \alpha_2((\overline{c_1} + kk_2) - \alpha_1 \underline{c_2})$
$\overline{p_0}$	$\overline{a_1(k_1)} - \alpha_1(\underline{a_2(k_2)} - \alpha_2 \overline{a_3}) = (\overline{c_0} + kk_1)\text{-}\alpha_1((\underline{c_1} + kk_2) - \alpha_2 \overline{c_2})$

Remainder (6) for ICP (5) and dominant polynomial (2) can be written as follows:

$$R = k \cdot k_0 - [\alpha] \cdot (([c_0] + k \cdot k_1) - [\alpha] \cdot (([c_1] + k \cdot k_2) - [\alpha] \cdot [c_2])) = 0. \tag{7}$$

3. General expressions for calculating dependent parameters of a controller

The text of your paper should be formatted as follows: Allocation area of a dominant pole is a range of a real axis of a complex plane, which left limit is α_1, right limit is α_2. Dominant polynomials for a left and a right limit of this range can be written as follows: $Q_1(s) = s + \alpha_1$ and $Q_2(s) = s + \alpha_2$.

Condition 1. Polynomial, which coefficients comply the following sequence of lower an upper limits of intervals $\underline{a_0}\,\overline{a_1}\,\underline{a_2}\,\overline{a_3}\ldots$, determines an allocation of real pole α_1. Angle of root locus branches departure from this pole is equal to 180°.

Condition 2. Polynomial, which coefficients comply the following sequence of lower an upper limits of intervals $\overline{a_0}\,\underline{a_1}\,\overline{a_2}\,\underline{a_3}\ldots$, determines an allocation of real pole α_2. Angle of root locus branches departure from this pole is equal to 0°.

Let us now apply (6) and conditions 1 and 2 for left and right limit of a dominant pole. The remainder *R* for $s = \alpha_1$ is determined by the following expression:

$$R_1 = \sum_{i=0}^{n} a_i \alpha_1^i (-1)^i, \tag{8}$$

where $a_i = \underline{a_i}$ *for* $i = 0, 2$ and $a_i = \overline{a_i}$ for $i = 1, 3$.

For $s = \alpha_2$

$$R_2 = \sum_{i=0}^{n} a_i \alpha_2^i (-1)^i, \tag{9}$$

where $a_i = \overline{a_i}$ for $i = 0, 2$ and $a_i = \underline{a_i}$ for $i = 1, 3$.

On a base of condition $R = 0$ and (7)-(9), a system of inequalities can be derived:

$$\begin{cases} R_1(\overline{k}) = k \cdot k_0 - \alpha_1 \cdot ((\overline{c_0} + k \cdot k_1) - \alpha_1 \cdot ((\underline{c_1} + k \cdot k_2) - \alpha_1 \cdot \overline{c_2})) = 0; \\ R_2(\overline{k}) = k \cdot k_0 - \alpha_2 \cdot ((\underline{c_0} + k \cdot k_1) - \alpha_2 \cdot ((\overline{c_1} + k \cdot k_2) - \alpha_2 \cdot \underline{c_2})) = 0. \end{cases} \tag{10}$$

Let us now express PID-controller parameter k_0 from each equation of (10)

IOP Conf. Series: Materials Science and Engineering **707** (2019) 012009 doi:10.1088/1757-899X/707/1/012009

$$\begin{cases} k_0(k_1,k_2) = \dfrac{\alpha_1 \cdot ((\overline{c_0} + k \cdot k_1) - \alpha_1 \cdot ((\underline{c_1} + k \cdot k_2) - \alpha_1 \cdot \overline{c_2}))}{k}; \\ k_0(k_1,k_2) = \dfrac{\alpha_2 \cdot ((\underline{c_0} + k \cdot k_1) - \alpha_2 \cdot ((\overline{c_1} + k \cdot k_2) - \alpha_2 \cdot \underline{c_2}))}{k}. \end{cases} \tag{11}$$

By equating right parts of equations (11) and expressing k_2 from the resulting expression, a function $k_2(k_1)$ can be found:

$$k_2(k_1) = \frac{\alpha_2 \cdot ((\underline{c_0} + k_1) - \alpha_2(\overline{c_1} - \alpha_2 \underline{c_2})) - \alpha_1 \cdot ((\overline{c_0} + k_1) - \alpha_1(\underline{c_1} - \alpha_1 \overline{c_2}))}{(\alpha_2^2 - \alpha_1^2) \cdot k^2}. \tag{12}$$

4. General expressions for placing non-dominant poles

Non-dominant polynomial of the considered system is a polynomial of the second order, so, there is a common expression for its roots. By finding them, let us derive expressions for values of PID-controller parameters, which provide desired allocation of non-dominant poles (real or complex-conjugate poles).

Roots of non-dominant polynomial can be calculated via following expression: $s_{1,2} = \dfrac{-p_1 \pm \sqrt{p_1^2 - 4p_0p_2}}{2p_2}$. Let us now formulate conditions of providing a desired allocation of non-dominant poles on the left from border η in two cases: if both of them are real and if they are complex-conjugate. In the first case, following inequalities must be complied:

$$\begin{cases} p_1^2 - 4p_0p_2 > 0; \\ \dfrac{-p_1 - \sqrt{p_1^2 - 4p_0p_2}}{2p_2} \le -\eta, \end{cases} \tag{13}$$

in the second case:

$$\begin{cases} p_1^2 - 4p_0p_2 < 0; \\ -\dfrac{p_1}{2p_2} < -\eta. \end{cases} \tag{14}$$

By rewriting (13) and (14), conditions of desired real non-dominant poles allocation can be derived:

$$\begin{cases} p_1^2 - 4p_0p_2 > 0; \\ p_0 \ge \eta p_1 - \eta^2 p_2, \end{cases} \tag{15}$$

for complex-conjugate poles:

$$\begin{cases} p_1^2 - 4p_0p_2 < 0; \\ p_1 > 2\eta^2 p_2. \end{cases} \tag{16}$$

Let us now expand conditions (15) and (16) for interval numbers. Condition (15) will have a following form:

$$\begin{cases} \underline{p_1^2(\vec{k})} - 4\overline{p_0(\vec{k})}\,\overline{p_2(\vec{k})} > 0; \\ \underline{p_0(\vec{k})} \ge \eta \overline{p_1(\vec{k})} - \eta^2 \underline{p_2(\vec{k})}. \end{cases} \tag{17}$$

Condition (16) will have a following form:

$$\begin{cases} \overline{p_1^2(\vec{k})} - 4\underline{p_0(\vec{k})}\,\underline{p_2(\vec{k})} < 0; \\ \underline{p_1(\vec{k})} > 2\eta \overline{p_2(\vec{k})}. \end{cases} \tag{18}$$

After substituting (12) and (13) to (17) or (18), a range of k_1 parameter values, which provide desired allocation of non-dominant poles, can be found.

5. Method of synthesizing a PID-controller for a third-order control system with interval parameters

1. On the base of transfer functions of a controller and a control object with interval parameters and a desired control quality, determined by a dominant pole $[\alpha]=[\alpha_1;\alpha_2]$, following function must be defined: $A(s)$ – is an ICP (1) of a closed-loop system; $Q(s)$ – is an interval dominant polynomial (2); $P(s)$ – non-dominant polynomial (3) with coefficients from Table 1; R – interval remainder (7).
2. On the base of (11) and (12) functions of controller parameters must be derived: $k_0(k_1,k_2)$, $k_2(k_1)$.
3. On the base of previously defined values of $[\alpha]$, η and coefficients of non-dominant polynomial, system of inequalities (17) or (18) must be derived.
4. Range of k_1 parameter values must be determined from the solution of inequalities system, derived in step 3.
5. Value of a k_1 parameter must be chosen; values of other parameters must be calculated with the help of functions $k_0(k_1,k_2)$ and $k_2(k_1)$, derived in step 2.

6. Example of controller synthesis

Control object of the considered system is described with a transfer function $W_{co}(s)=\dfrac{1}{c_2 s^2+c_1 s+c_0}$, where $c_0\in[1;3], c_1\in[0,5;0,7], c_2\in[0,03;0,05]$. Transfer function of a PID-controller is written as follows: $W_c(s)=\dfrac{k_2 s^2+k_1 s+k_0}{s}$. The aim of the synthesis is to allocate dominant pole of the system within the interval $[\alpha]=[1;1,5]$; non-dominant poles must be placed on the left from the line $\eta=4$.

According to the synthesis algorithm, an ICP of the considered system was derived: $[a_3]s^3+[a_2]s^2+[a_1]s+[a_0]=0$, where $a_0(k_0)=k_0$, $[a_1(k_1)]=[1;3]+k_1$, $[a_2(k_2)]=[0,5;0,7]+k_2$, $[a_3]=[0,03;0,05]$. Interval dominant polynomial has a following form: $Q(s)=s+[1;1,5]$; interval non-dominant polynomial is written as follows:

$$P(s,k_1,k_2)=[0,03;0,05]s^2+([0,425;0,67]+k_2)s+[-0,005;2,575]+k_1+[1;1,5]\cdot k_2,$$

interval remainder is determined as follows:

$$R(s,\vec{k})=k_0-[1;1,5]\cdot\left([1;1,5]\cdot([1;1,5]\cdot[0,03;0,05]-([0,5;0,7]+k_2))+([1;3]+k_1)\right).$$

Then, according to the second step of the synthesis algorithm, expressions (11) and (12) for the considered system were derived:

$$\begin{cases} k_0(k_1,k_2)=2.55+k_1-k_2; \\ k_0(k_1,k_2)=0.02625+1.5\cdot k_1-2.25\cdot k_2. \end{cases} \quad (19)$$

$$k_2(k_1)=0,4k_1-2,019 \quad (20)$$

Let us now synthesize the system with complex-conjugate non-dominant poles. By substituting (20) to (18), a set of k_1 values, which provide desired poles allocation, was found: $k_1\in[4,98;5,336]$. After defining k_1, values of k_0 and k_2 were calculated with the help of (19) and (20). It was determined, that

1. If $k_1=4.98$, $k_0=7.557$, $k_2=-0.027$, then $\eta=-4,032$.
2. If $k_1=5.336$, $k_0=7.771$, $k_2=0.115$, then $\eta=-5.45$.

Let us notice, that the right limit of the k_1 range of values corresponds to higher robust pole domination degree γ. In the figure 1, allocation areas of the synthesized system poles are shown. It is clear, that non-dominant poles are complex-conjugate.

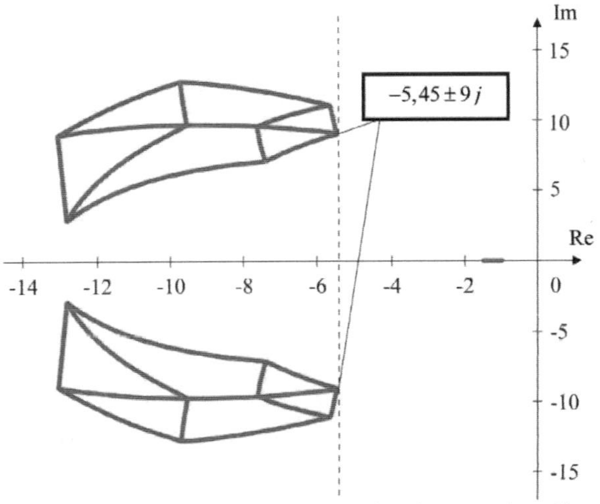

Figure 1. Areas of dominant and non-dominant poles allocation.

A step response of the synthesized system in the vertex $V = \underline{a_0}\,\overline{a_1}\,\underline{a_2}\,\overline{a_3}$, which corresponds to minimal stability degree, is shown in the figure 2. According to the figure 2, setting time is equal to 1.92 sec., overshoot is equal to 0%.

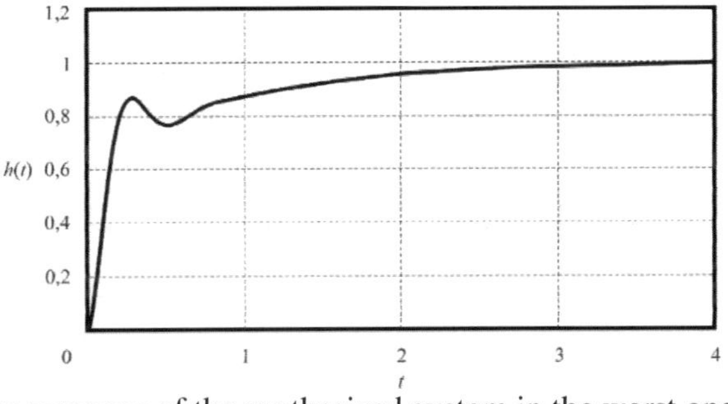

Figure 2. Step response of the synthesized system in the worst operating mode.

Let us now synthesize a system with real non-dominant poles. By substituting (20) to (17), a set of k_1 values, which provide desire poles allocation can be found: $k_1 \in [7{,}34; 7{,}4]$. It was determined, that

1. If $k_1 = 7{,}34$, $k_0 = 8{,}973$, $k_2 = 0{,}917$, then $\eta = -4{,}125$;
2. If $k_1 = 7{,}4$, $k_0 = 9.009$, $k_2 = 0{,}941$, then $\eta = -4{,}072$.

Let us notice, that the left limit of the k_1 range of values corresponds to higher robust pole domination degree γ. In the figure 3, allocation areas of the synthesized system poles are shown. It is clear, that non-dominant poles are real.

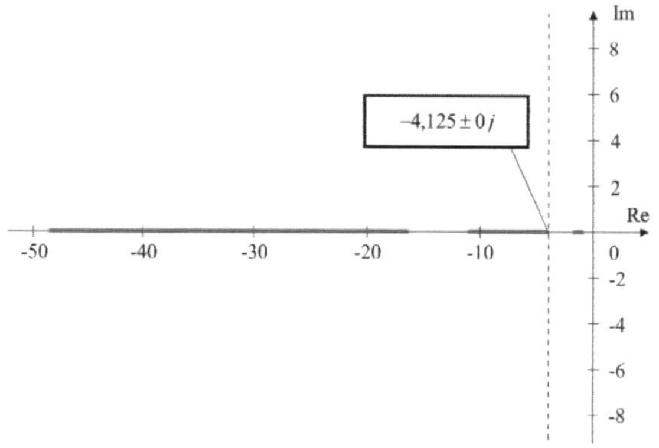

Figure 3. Areas of dominant and non-dominant poles allocation of the system with real poles.

A step response of the synthesized system in the vertex $V = \underline{a_0}\,\overline{a_1}\,\underline{a_2}\,\overline{a_3}$, which corresponds to minimal stability degree, is shown in the figure 4. According to the figure 4, setting time is equal to 1.9 sec., overshoot is equal to 0%.

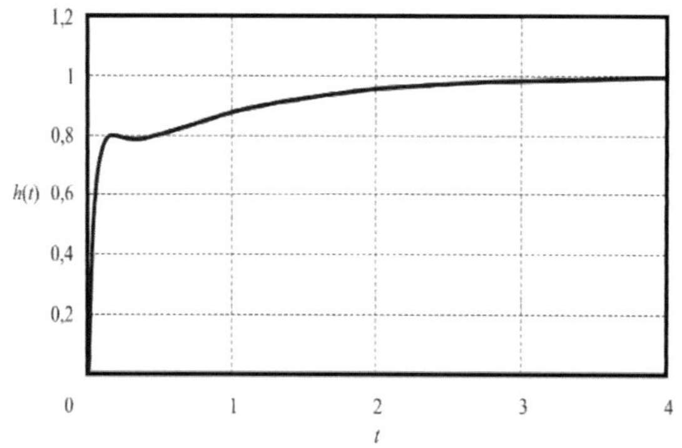

Figure 4. Step response of the synthesized system with the real poles.

7. Conclusion

The problem of parametric synthesis of a PID-controller, providing an aperiodic transient process of desired duration in a system of the third order with interval parameters, was solved. Developed synthesis method is based on a decomposition of an ICP in two polynomials: a dominant and a non-dominant one. Dominant polynomial determines an allocation of a dominant system pole, which, consequently, defines a quality of a transient process. Non-dominant polynomial allows placing non-dominant poles on a desired distance from the dominant pole.

To synthesize a controller, a set of conditions, linking controller parameters and interval coefficients of ICP, was derived. On the base of these conditions, a synthesis method was developed; an example of controller synthesis with the help of the method is provided.

8. Acknowledgments

The reported study is supported by the Ministry of Education and Science of Russian Federation (project #2.3649.2017/PCh).

9. References

[1] Ioannou P and Baldi P 2010 *Robust Adaptive Control: The Control Systems Handbook* (Boca Raton: CRC Press).

[2] Ioannou P and Sun J 2012 *Robust Adaptive Control* (United States: Courier Corporation).

[3] Anderson B, Jury E and Chaparro F 2014 Relations between real and complex polynomials for stability and aperiodicity conditions *Proc. of the 1974 IEEE Conf. on Decision and Control including 13th Symp. on Adaptive Processes (Phoenix)* 245–247.

[4] M. V. Meerov, E. I. Jury On aperiodicity robustness International Journal of Control, **70(2),** 193-201 (1998).
[5] Polyak B and Scherbakov P 2002 *Robust stability and control* (Moscow: Science).
[6] Bhattacharyya S, Chapellat H and Keel L 1995 *Robust control: The parametric approach* (United States: Prentice-Hall).
[7] Nesenchuk A 2002 Parametric synthesis of qualitative robust control systems using root locus fields. *Proc. of the 15th Triennial World Congress of The Int. Federation of Automatic Control (Barcelona)* 387-387.
[8] Nesenchuk A and Fedorovich S 2008 Parametric synthesis method of integral systems on the basis of root locus curves of Kharitonov polynomials *Autom. and Rem. Contr.* **69(7)** 1133-41.
[9] Nesenchuk A 2017 A method for synthesis of robust interval polynomials using the extended root locus. *Proc. of the American Control Conf. (Seattle)* 1715-25.
[10] Wang Y, Schinkel M and Hunt K 2002 PID and PID-like controller design by pole assignment within D-stable regions *Asian J. of Contr.* **4(4)** 423-32.
[11] Zhmud V, Dimitrov L and Yadrishnikov O 2014 Calculation of regulators for the problems of mechatronics by means of the numerical optimization method *Proc. of 12th Int. Conf. on Actual Problems of Electronic Instrument Engineering (Novosibirsk)* 739-744.
[12] Zhmud V, Reva I and Dimitrov L 2017 Design of robust systems by means of the numerical optimization with harmonic changing of the model parameters *J. of Phys.: Conf. Ser.* **803(1)**.
[13] Tagami T and Ikeda K 2003 Design of robust pole assignment based on Pareto-optimal solutions *Asian J. of Contr.* **5(2)** 195-205.
[14] Gayvoronskiy S, Ezangina T and Khozhaev I 2015 The interval-parametric synthesis of a linear controller of speed control system of a descent submersible vehicle *IOP Conf. Ser.: Mat. Sc. and Eng.* **93**.

IOP Conf. Series: Materials Science and Engineering **707** (2019) 012010 doi:10.1088/1757-899X/707/1/012010

Invariance, controlled invariance and conditioned invariance in structured systems and applications to disturbance decoupling

G Conte[1], A M Perdon[1], E Zattoni[2] and C H Moog[3]

[1] Dipartimento di Ingegneria dell'Informazione, Università Politecnica delle Marche, 60131 Ancona, Italy

[2] Department of Electrical, Electronic, and Information Engineering "G. Marconi", Alma Mater Studiorum Università di Bologna, 40136 Bologna, Italy

[3] Laboratoire des Sciences du Numérique de Nantes LS2N –UMR CNRS 6004, 44321 Nantes, France

E-mail: {gconte, a.m.perdon}@staff.univpm.it; elena.zattoni@unibo.it; Claude.Moog@ls2n.fr

Abstract. In this paper, dynamical systems whose structure is defined by means of a simple, directed graph are considered. These objects can be used to model structured systems or, more generally, networks of systems and systems of systems, where the relations between state, input and output variables or, respectively, between agents are known only for being zero or nonzero. Using an approach that is conceptually similar to the geometric approach developed for linear time-invariant systems, suitable notions of invariance, controlled invariance and conditioned invariance are introduced and related to the action of feedbacks. The results are used to provide general solvability conditions for disturbance decoupling problems expressed in graph-theoretic terms.

1. Introduction

Simple, directed graphs can be used to describe dynamical systems by associating the vertices to the state variables of the system and the edges to nonzero relationships between them. This representation can be used for modeling uncertain systems, whose equations contain parameters that are unknown except for being equal to or different from 0. Networks of systems and systems of systems can be represented analogously, associating the vertices of the graph to the agents of the network or to the components of a complex system, when the focus is on the influence that each agent or component exerts on the others, rather than on the individual dynamics of each of them. In all cases, we speak of structured systems to highlight the fact that the graph gives information only on those aspects of the dynamical structure of the system that do not depend on specific values of the nonzero parameters or on specific features of the communication lines that link the agents.

The study of linear structured systems was pioneered in [1] and it received contributions by many authors in the subsequent years. Investigation aims mainly at discovering and pointing out which properties of the system are determined by the structure of the graph and, therefore, can be said to be *structural* or, since they hold for any value of the nonzero parameters, *generic*.

Examples are structural (or generic) controllability and structural (or generic) observability. Classical control problems, like noninteractive control and disturbance decoupling, have been studied in that framework, together with problems related to the structure of finite zeros and zeros at infinity. An excellent survey of available results in the area, with an exhaustive list of references, can be found in [2].

More recently, graphs have been used to model the interaction of a plurality of independent agents in complex systems, like networked systems and, more generally, systems of systems (see [3], [4], [5], [6], [7], [8] and the references therein).

The approach to the study of structured systems we develop in this paper is based on geometric concepts and, as such, it is conceptually analogous to the structural geometric approach for classical linear systems of [9] and [10]. To do this, we restrict our attention to a particular class of structured systems, which is slightly less general than that considered e.g. in [2]. More specifically, we assume that, when the system is represented in state space form, each input variable appears in no more than one state equation. In terms of the underlying graph, this means that each input channel involves a single vertex. Structured systems of this kind are those considered in [11], [12], [13], those used in modelling quantum systems in [14] and those appearing in [15] and in examples in [3]. Such feature is crucial to introduce suitable notions of *invariance*, *controlled invariance* and *conditioned invariance* that can be naturally and consistently expressed in graph-theoretic terms. Controlled invariance is then related to and interpreted in terms of a notion that characterizes the (state or output) feedbacks whose action modifies the underlying graph of the structured system, called *essential feedback*. Also the notion of essential feedback can be naturally and simply expressed in graph-theoretic terms.

Invariance and feedback whose action modifies the underlying graph do not have a similar characterization for general structured systems, in which a single input variable may affect more than one state variable. This is probably why, in spite of the existence of a wide literature in this area, they have not, to the best of our knowledge, been previously introduced in graph-theoretic terms as we do here (compare with [2]). The definition of invariance and feedback in graph-theoretic terms for the considered class of structured systems is the main theoretic contribution of this paper.

The approach developed in this way can be applied to the study of a number of problems, which in particular include disturbance decoupling problems by means of state or output feedback. In regard to these problems, we are able to state necessary and sufficient solvability conditions for the considered class of structured systems in Theorem 1 and in Theorem 2. Such conditions are equivalent to those stated in Theorem 7 and 8 of [2] for general structured systems, but, thanks to the notions we introduce here, they are expressed in a more natural and simple way, which formally coincides with the classical formulation of solvability of the analogous problems in the framework of classical linear systems.

The paper is organizede as follows. In Section 2, we describe the graph representation of structured systems and we introduce the class of systems we consider. In Section 3, we introduce the fundamental notions of invariance, controlled invariance and conditioned invariance in graph-theoretic terms. Then, we define the notion of essential feedback and we show how it can be used to characterize controlled invariance. In Sections 4 and 5, we study the disturbance decoupling problem by means, respectively, of state feedback and of output feedback and we characterize solvability by means of necessary and sufficient conditions. Proofs of our results are only sketched and they will appear in complete form elsewhere. Section 6 contains conclusions and description of future work.

Notation Given two sets A and B, we will denote by $A\backslash B$ the set difference of A and B, that is the subset of A defined by $A\backslash B = \{a \in A, \ such \ that \ a \notin B\}$.

2. Preliminaries

Let $(G, \mathcal{E})$ be a simple directed graph (i.e. a directed graph without multiple edges and without auto-loops) with set of vertices $G = \{v_1, ..., v_n\}$ and set of edges $\mathcal{E} \subseteq G \times G$. If $(v_j, v_i) \in \mathcal{E}$, we say that v_j is the tail and v_i is the head of the edge (v_j, v_i). A path P in $(G, \mathcal{E})$ is an ordered finite sequence of edges $(e_1, ..., e_k)$ in which the head of the edge e_h coincides with the tail of the edge e_{h+1}. The tail of the first edge in a path P is called the tail of the path and the head of the last edge is called the head of the path.

Given a graph $(G, \mathcal{E})$ with n vertices, we associate to it an $n \times n$ matrix $A = [a_{ij}]$ whose entries are real, mutually independent parameters that satisfy the following conditions

- $a_{ij} \neq 0$ for $i \neq j$ if and only if $(v_j, v_i) \in \mathcal{E}$ (i.e. there is an edge from v_j to v_i in $(G, \mathcal{E})$).

Note that no condition is imposed on a_{ij} for $i = j$.

Letting $G_{in} = \{v_{i_1}, ..., v_{i_m}\} \subseteq G$ be a subset of vertices, we associate to the pair $((G, \mathcal{E}), G_{in})$ an $n \times m$ matrix $B = [b_{ij}]$ whose entries are real, mutually independent parameters that satisfy the following conditions

- $b_{ij} = 1$ (or more generally $b_{ij} \neq 0$) if $v_{i_j} = v_i$ (that is: if the j-th element of G_{in} is equal to v_i)
- $b_{ij} = 0$ otherwise.

Note that in any column of B there is just one entry different from 0, while in any row of B there is at most one entry different from 0.

Letting $G_{out} = \{v_{i_1}, ..., v_{i_p}\} \subseteq G$ be a subset of vertices, we associate to the pair $((G, \mathcal{E}), G_{out})$ a $p \times n$ matrix $C = [c_{ij}]$ whose entries are real, mutually independent parameters that satisfy the following conditions

- $c_{ij} = 1$ (or more generally $c_{ij} \neq 0$) if $v_{i_j} = v_i$ (that is: if the j-th element of G_{out} is equal to v_i)
- $c_{ij} = 0$ otherwise.

Note that in any row of C there is just one entry different from 0, while in any column of C there is at most one entry different from 0.

In representing graphically the triple $((G, \mathcal{E}), G_{in}, G_{out}))$, we use arrows between vertices to indicate edges and we mark the elements of G_{in} by ingoing arrows and the elements of G_{out} by outgoing arrows, as in Figure 1.

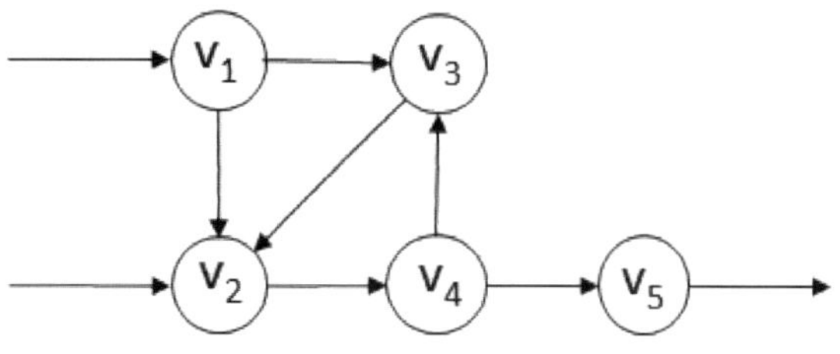

Figure 1. A simple, directed graph with $G_{in} = \{v_1, v_2\}$ and $G_{out} = \{v_5\}$.

The **structured system** Σ associated to the triple $((G, \mathcal{E}), G_{in}, G_{out}))$ is the linear time-invariant system described in parametric state space form by the equation

$$\Sigma \equiv \begin{cases} \dot{x}(t) &= Ax(t) + Bu(t) \\ y(t) &= Cx(t) \end{cases} \quad (1)$$

with state $x \in \mathbb{R}^n$, input $u \in \mathbb{R}^m$ and output $y \in \mathbb{R}^p$.

Structured systems are used to model families of linear systems whose elements are characterized by specific values of the parameters appearing in (1), or uncertain systems where parameters account for unknown relations between variables. In this case, the vertex $v_i \in G$ corresponds to the component x_i of the state vector $x = (x_1, ..., x_n)^\top$ of Σ. The elements of G_{in} identify the components of the state vector whose dynamics is directly influenced by the input, while the elements of G_{out} identify the component of the state that are directly measurable from the output. One can also interpret Σ as the model of a network of systems, or dynamical agents, or a complex system of systems. In this case, the vertices of the graph represent systems, or agents, in the network and the edges represent monodirectional communication lines, whose weights are expressed by the parameters, that account for the way in which agents influence each other. Vertices in G_{in} correspond to agents that can be influenced by inputs coming from the external environment and vertices in G_{out} corresponds to agents that send outputs to the external environment.

In all the above mentioned interpretations, the topology of the underlying graph can be exploited to obtain information on dynamical properties that are shared by all the elements in the family, i.e. that are valid for all the values of the uncertain parameters, or that characterize the overall properties of the network.

Note that, as pointed out above, the graph gives no information about the diagonal elements a_{ii} of A since it does not contain auto-loops. Actually, the fact that an element of that kind is equal to or different from 0 does not affect the dynamical properties we will deal with in the sequel.

Remark 1 *It is important to remark that the class of structured systems we consider is less general than that considered in [2]. The particular structure of the matrix B in our framework, in fact, means that each input variable in (1) appears in only one state equation or, in other words, that it influences directly the dynamics of only one state variable. This restriction is relevant for assuring validity and consistency to the notions that will be introduced in the following section and for deriving the results that follow. Structured systems of this class are explicitely considered in [11], [12], [13], they are used in modelling quantum systems [14] and example appears in [15], [3]. Similarly, the particular structure of the matrix C means that each output variable coincides, except for a scale factor, with one state variable.*

Notation We denote the structured system associated, as described above, to the triple $((G, \mathcal{E}), G_{in}, G_{out})$ by $\Sigma((G, \mathcal{E}), G_{in}, G_{out})$ or simply by Σ, if no confusion arises.

3. Invariance and controlled invariance

Given a structured system $\Sigma((G, \mathcal{E}), G_{in}, G_{out})$, we call *dynamically closed* a subset of vertices $V \subseteq G$ if and only if any path P whose tail and head are in V consists of edges whose tails and heads are in V.

A subsystem of $\Sigma((G, \mathcal{E}), G_{in}, G_{out})$ is a structured system $\Sigma((V, \mathcal{E}_{|V}), V_{in}, V_{out})$ such that

- V is a dynamically closed subset of G,
- $E_{|V} = \{(v_j, v_i) \in \mathcal{E},\ \textit{such that}\ v_j \in V\ \textit{and}\ v_i \in V\}$,
- $V_{in} = V \cap G_{in}$,
- $V_{out} = V \cap G_{out}$.

In accordance, from a conceptual point of view, with the geometric approach developed for linear systems in [9] and [10], we can introduce in our framework the notions of *invariance* and *controlled invariance* as follows.

Definition 1 *Given a structured system $\Sigma((G,\mathcal{E}),G_{in},G_{out})$, a subset $V \subseteq G$ of vertices is said to be* ***invariant*** *for Σ if $(v_j,v_i) \in \mathcal{E}$ with $v_j \in V$ implies $v_i \in V$.*

If $V \subseteq G$ is invariant, then it is dynamically closed and $\Sigma((V,\mathcal{E}_{|V}),V_{in},V_{out})$ is a subsystem of $\Sigma((G,\mathcal{E}),G_{in},G_{out})$.

Definition 2 *Given a structured system $\Sigma((G,\mathcal{E}),G_{in},G_{out})$, a subset $V \subseteq G$ of vertices is said to be* ***controlled invariant*** *for Σ if $(v_j,v_i) \in \mathcal{E}$ with $v_j \in V$ implies $v_i \in V \cup G_{in}$.*

Given a structured system $\Sigma((G,\mathcal{E}),G_{in},G_{out})$, we are interested in considering state feedbacks that modify the dynamics expressed by the underlying graph. Remarking that the action of a state feedback may only involve the relationship between any component of the state (or any agent) and the components (agents) whose dynamics is directly affected by the inputs, we fix our attention on the subset $\mathcal{E}_{in} \subseteq \mathcal{E}$ defined by

$$\mathcal{E}_{in} = \{(v_j,v_i) \in \mathcal{E}, \text{ such that } v_i \in G_{in}\}$$

and we introduce the following definition.

Definition 3 *Given a structured system $\Sigma((G,\mathcal{E}),G_{in},G_{out})$, an* ***essential state feedback*** *consists of a subset $\mathcal{F} \subset G \times G_{in}$. The action of an essential state feedback $\mathcal{F}$ on the structured system $\Sigma((G,\mathcal{E}),G_{in},G_{out})$ gives rise to the compensated system $\Sigma((G,\mathcal{E}^{\mathcal{F}}),G_{in},G_{out})$, where $\mathcal{E}^{\mathcal{F}} = (\mathcal{E}\cup\mathcal{F})\backslash(\mathcal{E}\cap\mathcal{F}) = \{(v_j,v_i) \in \mathcal{E}\cup\mathcal{F}$ such that $(v_j,v_i) \notin \mathcal{E}\cap\mathcal{F}\}$.*

Note that applying an essential state feedback to $\Sigma((G,\mathcal{E}),G_{in},G_{out})$ means to modify $\mathcal{E}_{in}$, either by adding new elements of the form (v_j,v_i) with $v_j \in G$ and $v_i \in G_{in}$ to it or by removing elements of the form (v_j,v_i) with $v_i \in G_{in}$, if present, from it. We can associate to the essential state feeback $\mathcal{F}$ the relation $u = Fx$, where $F = [f_{ij}]$ is an $m \times n$ matrix whose entries are real, mutually independent parameters. To describe how to do this, let us first remark that if (v_j,v_i) belongs to $\mathcal{E}\cap\mathcal{F}$, then we have that v_i is, for some k, the k-element of G_{in}, i.e. $v_i = v_{i_k} \in G_{in}$, and, hence, $b_{ik} \neq 0$ in the matrix B. Now, let us take the parameters f_{ij} in such a way that they satisfy the following conditions

- $f_{ij} \neq 0$ if and only if $(v_j,v_i) \in \mathcal{F}$
- $f_{ij} = -a_{ij}/b_{ij}$ if $(v_j,v_i) \in \mathcal{F}$ with $v_i = v_{i_k} \in G_{in}$ and $a_{ik} \neq 0$.

Note that no condition is imposed on f_{ij} if $i = j$ and that only the first of the two conditions above applies if, for $i \neq j$ and $(v_j,v_i) \in \mathcal{F}$ with $v_i = v_{i_k} \in G_{in}$, one has $a_{ik} = 0$.

With the above choice, the compensated system $\Sigma((G,\mathcal{E}^{\mathcal{F}}),G_{in},G_{out})$ turns out to be defined in parametric form by the following set of equations

$$\Sigma^F \equiv \begin{cases} \dot{x}(t) &= (A+BF)x(t) + Bu(t) \\ y(t) &= Cx(t) \end{cases} \tag{2}$$

Example 1 *Let us consider the graph of Figure 1 and the associated structured system $\Sigma = ((G,\mathcal{E}),G_{in},G_{out})$, which is described in parametric state space form by the following equations:*

$$\Sigma \equiv \begin{cases} \dot{x}_1(t) &= a_{11}x_1 + b_{11}u_1 \\ \dot{x}_2(t) &= a_{21}x_1 + a_{22}x_2 + a_{23}x_3 + b_{22}u_2 \\ \dot{x}_3(t) &= a_{31}x_1 + a_{33}x_3 + a_{34}x_4 \\ \dot{x}_4(t) &= a_{42}x_2 + a_{44}x_4 \\ \dot{x}_5(t) &= a_{54}x_4 + a_{55}x_5 \\ y(t) &= c_{15}x_5(t) \end{cases} \tag{3}$$

A generic state feedback $u = Fx$ is given by the equations

$$\begin{cases} u_1(t) = f_{11}x_1 + f_{12}x_2 + f_{13}x_3 + f_{14}x_4 + f_{15}x_5 \\ u_2(t) = f_{21}x_1 + f_{22}x_2 + f_{23}x_3 + f_{24}x_4 + f_{25}x_5 \end{cases} \tag{4}$$

where the terms f_{ij} are real, independent parameters. Applying such state feedback to Σ, the resulting compensated system turns out to be described by a triple $((G, \mathcal{E}'), G_{in}, G_{out})$ that may differ from $((G, \mathcal{E}), G_{in}, G_{out})$ only because edges have been added to or removed from $\mathcal{E}$. More precisely, edges are added if some of the parameters $f_{12}, f_{13}, f_{14}, f_{15}, f_{24}, f_{25}$ is different from 0, while edges are removed if $f_{21} = -a_{21}/b_{22}$ or $f_{23} = -a_{23}/b_{22}$. The values of the remaining parameters in (4) do not affect $\mathcal{E}$. Note that, for instance, if $\mathcal{F}$ is the essential state feedback defined by $\mathcal{F} = \{(v_1, v_2), (v_3, v_2), (v_5, v_2)\}$, its action on $\Sigma = ((G, \mathcal{E}), G_{in}, G_{out})$ produces the compensated system $\Sigma((G, \mathcal{E}^{\mathcal{F}}), G_{in}, G_{out})$ that is described by the graph in Figure 2.

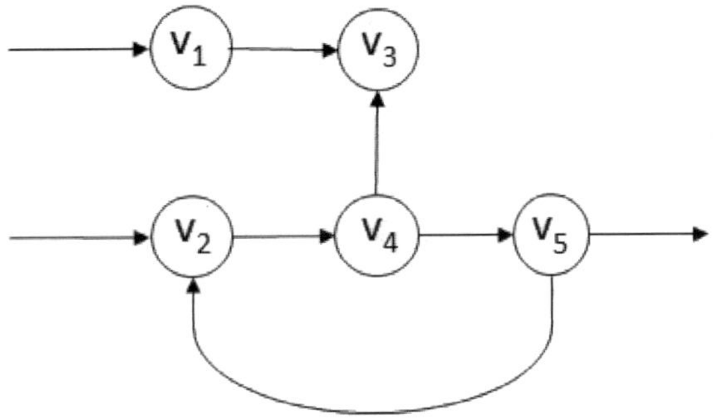

Figure 2. The structured compensated system $\Sigma((G, \mathcal{E}^{\mathcal{F}}), G_{in}, G_{out})$.

We can now give in our framework a simple but fundamental characterization of controlled invariance that is analogous to the one given for the analogous concept in the classical linear framework (compare with [9], [10]).

Proposition 1 *Given a structured system $\Sigma((G, \mathcal{E}), G_{in}, G_{out})$, a subset $V \subseteq G$ of vertices is controlled invariant for Σ if and only if there exists an essential state feedback $\mathcal{F}$ such that V is invariant for the compensated system $\Sigma((G, \mathcal{E}^{\mathcal{F}}), G_{in}, G_{out})$.*

Hint of proof. Assume that $V \subseteq G$ is controlled invariant for Σ and take $\mathcal{F} = \{(v_j, v_i) \in \mathcal{E}$, *such that* $v_j \in V$ *and* $v_i \in G_{in}\}$. We have that $\mathcal{E}^{\mathcal{F}} = (\mathcal{E} \cup \mathcal{F}) \backslash (\mathcal{E} \cap \mathcal{F}) = \mathcal{E} \backslash \mathcal{F}$ and, therefore, $(v_j, v_i) \in \mathcal{E}^{\mathcal{F}}$ with $v_j \in V$ implies $v_i \in V$. Conversely, if V is invariant for $\Sigma((G, \mathcal{E}^{\mathcal{F}}), G_{in}, G_{out})$ for some feedback $\mathcal{F}$, we have that $(v_j, v_i) \in \mathcal{E}$ with $v_j \in V$ implies that either $v_i \in V$ or $(v_j, v_i) \in \mathcal{F}$. Therefore, $(v_j, v_i) \in \mathcal{E}$ with $v_j \in V$ implies that either $v_i \in V$ or $v_i \in G_{in}$.
The essential state feedback $\mathcal{F}$ constructed in the proof of Proposition 1 is not the only one that makes V invariant in the compensated system. Let $\mathcal{E}_1 \subseteq \mathcal{E}$ be defined by

$$\mathcal{E}_1 = \{(v_j, v_i) \in \mathcal{E},\ \textit{such that } v_i \in G_{in} \textit{ and } v_j \in V \textit{ implies } v_i \in V \cap G_{in}\}.$$

Any feedback $\mathcal{F}'$ that satisfies the condition

$$\mathcal{F} \subseteq \mathcal{F}' \subseteq \mathcal{E}_1 \tag{5}$$

makes V invariant in the compensated system $\Sigma((G, \mathcal{E}^{\mathcal{F}'}), G_{in}, G_{out})$. Any feedback $\mathcal{F}$ that has the property of making V invariant in $\Sigma((G, \mathcal{E}^{\mathcal{F}}), G_{in}, G_{out})$ is called a *friend* of V.

Proposition 2 *Given a structured system $\Sigma((G, \mathcal{E}), G_{in}, G_{out})$ and a subset $K \subseteq G$, there exists a maximal subset of vertices V such that $V \subseteq K$ and V is controlled invariant for Σ. We denote such subset by $V^*(\mathcal{E}, G_{in}, K)$ or simply by V^* if no confusion arises.*

Given a structured system $\Sigma((G, \mathcal{E}), G_{in}, G_{out})$ and a subset $K \subseteq G$, it is possible to construct $V^*(\mathcal{E}, G_{in}, K)$ by considering the sequence of subset $V_k \subseteq G$ defined recursively by

$$\begin{array}{rcl} V_0 & = & K \\ V_{k+1} & = & V_k \backslash \{v_j,\ such\ that\ (v_j, v_i) \in \mathcal{E}\ and\ v_i \notin V_k \cup G_{in}\}. \end{array} \tag{6}$$

Clearly, V_k converges to $V^*(\mathcal{E}, G_{in}, K)$ (which may be empty) in at most r steps, where r is the cardinality of K.

4. Disturbance decoupling by state feedback

A structured system subject to a disturbance is a system $\Sigma_D((G, \mathcal{E}), G_{in}, G_{out})$ in which G_{in} is partitioned as $G_{in} = G_c \cup G_d$ (possibly with $G_c \cap G_d = \emptyset$) and it is assumed that the disturbance inputs act directly on the dynamics of the variables (or agents) corresponding to the vertices in G_d, while the control input act directly on the dynamics of the variables (or agents) corresponding to the vertices in G_c. In that situation, we write $\Sigma((G, \mathcal{E}), (G_c \cup G_d), G_{out})$ and we can consider the following problem.

Problem 1 *Given a disturbed structured system $\Sigma((G, \mathcal{E}), (G_c \cup G_d), G_{out})$, the* ***Disturbance Decoupling Problem by State Feedback*** *(DDPSF) consists in finding an essential state feedback $\mathcal{F}$, if any exists, such that in the compensated system $\Sigma((G, \mathcal{E}^{\mathcal{F}}), (G_c \cup G_d), G_{out})$ there is no path in $(G, \mathcal{E}^{\mathcal{F}})$ with tail in G_d and head in G_{out} (this means that the output is not influenced by the disturbance).*

Applying the procedure that has been used to derive the system of equations (1), we get the following representation in parametric terms of the disturbed system $\Sigma((G, \mathcal{E}), (G_c \cup G_d), G_{out})$

$$\Sigma_D \equiv \left\{ \begin{array}{rcl} \dot{x}(t) & = & Ax(t) + Bu(t) + Dd(t) \\ y(t) & = & Cx(t) \end{array} \right. \tag{7}$$

where, the matrices $B = [b_{ij}]$ and $D = [d_{ij}]$ have dimensions, respectively, $n \times m_1$ and $n \times m_2$, with $m_1 = card(G_c)$ and $m_2 = card(G_d)$ and, letting $G_c = \{v_{c_1}, ..., v_{c_{m_1}}\}$ and $G_d = \{v_{d_1}, ..., v_{d_{m_2}}\}$, their entries satisfies the following conditions

- $b_{ij} = 1$ (or more generally $b_{ij} \neq 0$) if $v_{c_j} = v_i \in$ (that is: if the j-th element of G_c is equal to v_i)
- $d_{ij} = 1$ (or more generally $b_{ij} \neq 0$) if $v_{d_j} = v_i \in$ (that is: if the j-th element of G_d is equal to v_i)
- $b_{ij} = 0$ and $d_{ij} = 0$ otherwise

with control input $u \in \mathbb{R}^{m_1}$ and disturbance input $d \in \mathbb{R}^{m_2}$.

Solvability of the DDPSF stated above means solvability of the same problem for all values of the parameters which appear in (7). We can say that solvability of the DDPSF for $\Sigma((G, \mathcal{E}), (G_c \cup G_d), G_{out})$ means **structural solvability** of the problem. The following theorem gives necessary and sufficient condition for the solvability of the DDPSF and it also indicates how to construct a feedback $\mathcal{F}$ that solves it, if any exists.

Theorem 1 *Given a disturbed system $\Sigma((G, \mathcal{E}), (G_c \cup G_d), G_{out})$, let $K = G \backslash G_{out}$. Then, the associated DDPSF is solvable if and only if the condition*

$$G_d \subseteq V^*(\mathcal{E}, G_c, K). \tag{8}$$

is satisfied.

Hint of proof For sufficiency, let $\mathcal{F}$ be a friend of $V^*(\mathcal{E}, G_c, K)$, so that $V^*(\mathcal{E}, G_c, K)$ is invariant in $\Sigma((G, \mathcal{E}^{\mathcal{F}}), (G_c \cup G_d), G_{out})$. Condition (8) implies that any edge in $\mathcal{E}^{\mathcal{F}}$ with tail in G_d has its head in $V^*(\mathcal{E}, G_c, K)$ and, therefore, in K. This implies that there are no paths in $(G, \mathcal{E}^{\mathcal{F}})$ with tail in G_d and head in G_{out}.
For necessity, let $\mathcal{F}$ be a solution and consider the largest invariant $V \subseteq K$ for the compensated system $\Sigma((G, \mathcal{E}^{\mathcal{F}}), (G_c \cup G_d), G_{out})$. Clearly, $G_d \subseteq V$ and V is controlled invariant for $\Sigma((G, \mathcal{E}), (G_c \cup G_d), G_{out})$. The conclusion follows by maximality of $V^*(\mathcal{E}, G_c, K)$

Remark 2 *Comparing the conditions of Theorem 1 with those stated in Theorem 7 of [2] for general structured systems, it appears clearly that the notions of controlled invariance and essential feedback introduced in graph theoretic terms in the previous section simplify the situation and allow a quite natural solution of the DDPSF.*

Example 2 *Let us consider the structured disturbed system $\Sigma((G, \mathcal{E}), (G_c \cup G_d), G_{out})$, where $(G, \mathcal{E})$ is the graph of Figure 1 and $G_c = \{v_2\}$, $G_d = \{v_1\}$. We have that $K = \{v_1, v_2, v_3, v_4\}$ and computation shows that $V^*(\mathcal{E}, G_c, K) = \{v_1, v_3\}$. Applying the essential state feedback $\mathcal{F}$ defined by $\mathcal{F} = \{(v_1, v_2), (v_3, v_2)\}$, which is a friend of V^*, we get the compensated system $\Sigma((V, E^{\mathcal{F}}), (G_c \cup G_d), G_{out})$ described by the graph of Figure 3, whose output is not influenced by the disturbance that acts on the dynamics of the component (agent) corresponding to v_1.*

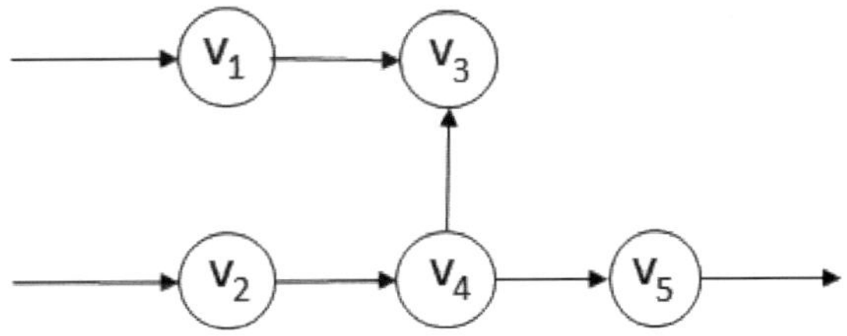

Figure 3. The conpensated system $\Sigma((V, E^{\mathcal{F}}), G_c \cup G_d, G_{out})$.

5. Disturbance decoupling by output feedback

Given a structured system $\Sigma((G, \mathcal{E}), G_{in}, G_{out})$, together with that of essential state feedback, we can consider a notion of essential output feedback. In order to state it formally, let us consider the subset $\mathcal{E}_{oi} \subseteq \mathcal{E}$ defined by

$$\mathcal{E}_{oi} = \{(v_j, v_i) \in \mathcal{E},\ \textit{such that } v_j \in G_{out} \textit{ and } v_i \in G_{in}\}.$$

Then, we introduce the following definition.

Definition 4 *Given a structured system $\Sigma((G, \mathcal{E}), G_{in}, G_{out})$, an* ***essential output feedback*** *consists of a subset $\mathcal{F} \subset G_{out} \times G_{in}$. The action of an essential output feedback $\mathcal{F}$ on the structured system $\Sigma((G, \mathcal{E}), G_{in}, G_{out})$ gives rise to the compensated structured system $\Sigma((G, \mathcal{E}^{\mathcal{F}}), G_{in}, G_{out})$, where $\mathcal{E}^{\mathcal{F}} = (\mathcal{E} \cup \mathcal{F}) \backslash (\mathcal{E} \cap \mathcal{F}) = \{(v_j, v_i) \in \mathcal{E} \cup \mathcal{F}$ such that $(v_j, v_i) \notin \mathcal{E} \cap \mathcal{F}\}$.*

In the present framework, output feedbacks form a restricted class of state feedbacks. Note that applying an essential output feedback to $\Sigma((G, \mathcal{E}), G_{in}, G_{out})$ means to modify $\mathcal{E}_{oi}$, which is contained in $\mathcal{E}_i$, either by adding new elements of the form (v_j, v_i) with $v_j \in G_{out}$ and $v_i \in G_{in}$ to it or by removing elements of the form (v_j, v_i) with $v_i \in G_{in}$, if present, from it.
Now, given a disturbed system $\Sigma((G, \mathcal{E}), (G_c \cup G_d), G_{out})$, in which G_{out} is partitioned as

$G_{out} = G_{co} \cup G_{mo}$, we can consider the problem of decoupling the disturbance from the (controlled) output variables corresponding to G_{co} by means of an output feedback that exploits the (measured) output variables corresponding to G_{mo}.

Problem 2 *Given a disturbed structured system $\Sigma((G,\mathcal{E}),(G_c \cup G_d),(G_{co} \cup G_{mo})$, the* ***Disturbance Decoupling Problem by Output Feedback*** *(DDPOF) consists in finding a essential output feedback $\mathcal{F} \subseteq G_{co} \times G_c$, if any exists, such that in the structured system $\Sigma((G,\mathcal{E}^{\mathcal{F}}),(G_c \cup G_d),(G_{co} \cup G_{mo})$ there is no path in $(G,\mathcal{E}^{\mathcal{F}})$ with tail in G_d and head in G_{co} (this means that the controlled output is not influenced by the disturbance).*

In order to provide solvability conditions for the DDPOF, it is convenient to introduce in our framework the notion of conditioned invariance. This is done as follows.

Definition 5 *Given a structured system $\Sigma((G,\mathcal{E}),G_{in},G_{out})$, a subset $S \subseteq G$ of vertices is said to be* ***conditioned invariant*** *for Σ if $(v_j,v_i) \in \mathcal{E}$ with $v_j \in S \cap G \backslash G_{out}$ implies $v_i \in S$.*

Conditioned invariant subsets have a number of structural properties that are relevant for our study.

Proposition 3 *Given a structured system $\Sigma((G,\mathcal{E}),G_{in},G_{out})$ and a subset $U \subseteq G$, there exists a minimal subset of vertices S such that $S \supseteq U$ and S is conditioned invariant for Σ. We denote such subset by $S^*(\mathcal{E},G_{out},U)$ or simply by S^* if no confusion arises.*

Given a structured system $\Sigma((G,\mathcal{E}),G_{in},G_{out})$ and a subset $U \subseteq G$, it is possible to construct $S^*(\mathcal{E},G_{out},U)$ by considering the sequence of subset $S_k \subseteq G$ defined recursively by

$$\begin{array}{rcl} S_0 & = & U \\ S_{k+1} & = & S_k \cup \{v_i, \text{ such that } (v_j,v_i) \in E \text{ and } v_j \in S_k \cap G_{out}\}. \end{array} \tag{9}$$

Clearly, S_k converges to $S^*(\mathcal{E},G_{out},K)$ (which may coincide with G) in at most $n-r$ steps, where r is the cardinality of U.
Now, we can state the following result.

Theorem 2 *Given a disturbed system $\Sigma((G,\mathcal{E}),(G_c \cup G_d),(G_{co} \cup G_{mo}))$, let $K = V \backslash G_{co}$. Then, the associated DDPOF is solvable if and only if the condition*

$$S^*(\mathcal{E},G_{mo},G_d) \subseteq V^*(\mathcal{E},G_c,K). \tag{10}$$

is satisfied.

Hint of proof Condition (10) is equivalent to the fact that any path in $(G,\mathcal{E})$ with tail in G_d and head in G_{co} contains at least one edge (v_j,v_i) with tail $v_j \in G_{mo}$ and head $v_i \in V_i$. Then, the solutions of the problem are the essential output feedbacks that remove all edges of that kind.

Remark 3 *Also in this case, as in that of the DDPSF, comparing the conditions of Theorem 2 with those stated in Theorem 8 of [2] for general structured systems, it appears clearly that the notions of controlled invariance and essential feedback introduced in graph theoretic terms in the Section 3 simplify the situation and allow a quite natural solution of the DDPOF.*

Example 3 *Consider the very simple example provided by the structured, disturbed system $\Sigma((G,\mathcal{E}),(G_c \cup G_d),(G_{co} \cup G_{mo}))$ described by the graph of Figure 4 on the left, where $G_c = \{v_c\}$, $G_d = \{v_d\}$, $G_{co} = \{v_{co}\}$, $G_{mo} = \{v_{mo}\}$ and the associated DDPOF. Clearly,*

$$S^*(\mathcal{E},G_{mo},G_d) = G_d = \{v_d\} \subseteq \{v_d,v_{mo}\} = V^*(\mathcal{E},G_c,K)$$

and condition (10) is satisfied. The essential output feedback $\mathcal{F} = \{(v_{mo},v_c)\}$ gives rise to the compensated system $\Sigma((G,\mathcal{E}^{\mathcal{F}}),(G_c \cup G_d),(G_{co} \cup G_{mo}))$, whose underlying graph is illustrated in Figure 4 on the right, and it solves the DDPOF.

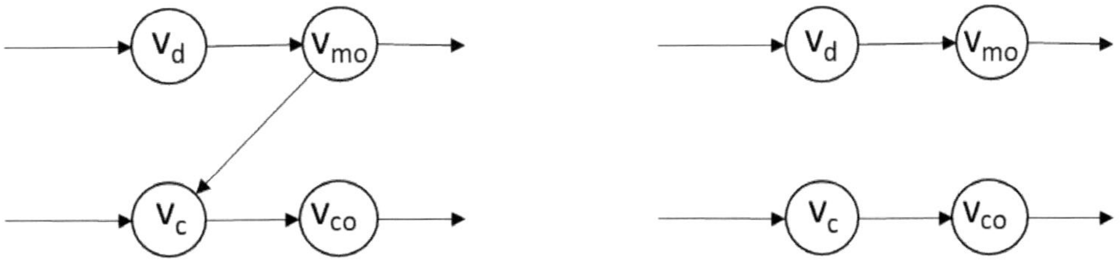

Figure 4. The structured stystem $\Sigma((G,\mathcal{E}),(G_c \cup G_d),(G_{co} \cup G_{mo}))$ (on the left) and the compensated stystem $\Sigma((G,\mathcal{E}^{\mathcal{F}}),(G_c \cup G_d),(G_{co} \cup G_{mo}))$ (on the right).

6. Conclusions

A novel approach to the study of a class of structured systems has been developed introducing and using novel graph-theoretic notions of invariance, controlled invariance, conditioned invariance and essential feedback. It has been shown that this approach provide a natural interpretation of relevant properties of the dynamics and that it can be efficiently used to characterize, in a new simple way, solvability of classical disturbance decoupling problems. Future work will aim at exploiting this approach in other noninteracting control problems for structured systems and in developing specific applications to networked systems and systems of systems.

7. References

[1] Lin C. T. (1974). *Structural controllability*, IEEE Transactions on Automatic Control, AC-19, 201–208
[2] Dion J.M., Commault C., van der Woude J. (2003). *Generic properties and control of linear structured systems: a survey*, Automatica 39, 1125–1144
[3] Mesbahi M., Egerstedt M. (2010). *Graph Theoretic Methods in Multiagent Networks*, Princeton University Press
[4] Chapman A., Mesbahi M. (2013). *On strong structural controllability of networked systems: A constrained matching approach.* American Control Conference (ACC), 61266131
[5] Gao J., Liu Y. Y., DSouza R. M., Barabasi A. L. (2014). *Target control of complex networks.* Nature Communications, 5
[6] Liu Y.Y., Barabasi A.L.(2016). *Control principles of complex systems*, Rev. Mod. Phys., 88
[7] Rahmani A., Ji M., Mesbahi M., Egerstedt M. (2009). *Controllability of multi-agent stystems from a graph-theoretic perspective*, SIAM J. Control Optim., Vol. 48, No. 1, pp. 162-186
[8] Harrison W. K. (2016). *The Role of Graph Theory in System of Systems Engineering*, IEEE Access, 4, 1716–1742
[9] Basile G., Marro G. (1992). *Controlled and Conditioned Invariants in Linear System Theor*, Prentice Hall, Englewood Clffs, New Jersey
[10] Wonham M. (1985). *Linear Multivariable Control: A Geometric Approach*, 3rd ed., Springer-Verlag, New York
[11] van Waarde H.J., Monshizadeh N., Trentelman H.L., Camlibel M.K. (2019). *Strong Structural Controllability and Zero Forcing*, in Structural Methods in the Study of Complex Systems, E. Zattoni, A. M. Perdon and G. Conte Eds., Springer Lectures Notes in Control and Information Science, 482, 92–114
[12] Trefois M., Delvenne J. C. (2015). *Zero forcing number, constrained matchings and strong structuralcontrollability*, Linear Algebra and its Applications 484, 199–218
[13] Monshizadeh N., Zhang S., Camlibel M.K. (2014). *Zero forcing sets and controllability of dynamical systems defined on graphs*, IEEE Transactions on Automatic Control 59(9), 2562–2567
[14] Burgarth D., DAlessandro D., Hogben L., Severini S., Young M. (2013). *Zero forcing, linear and quantum controllability for systems evolving on networks*, IEEE Transactions on Automatic Control 58, 2349–2354
[15] Califano C., Scharbarg E. Magdelaine N., Moog C. H. (2018). *Diabetic Gastroparesis Modeling and Observer Design* IFAC PapersOnLine 51-13 (2018), 97–102

IOP Conf. Series: Materials Science and Engineering **707** (2019) 012011 doi:10.1088/1757-899X/707/1/012011

Model predictive power distribution in an electric vehicle with a fuel cell

Soeren Scherler[1], **Xiaobo Liu-Henke**[1]

[1]Ostfalia University of Applied Sciences, Department of Mechanical Engineering, Institute for Mechatronics, Salzdahlumer Str. 46/48, 38302 Wolfenbuettel, Germany

E-Mail: so.scherler@ostfalia.de

Abstract. Within the subproject "Intelligent Electric Vehicles with Range Extender in Traffic Systems with Vehicle 4.0" (iREX 4.0), which is part of the joint project "Future Vehicle Technologies in the Open Region Lab" ("Zukünftige Fahrzeugtechnologien im Open Region Lab", ZuFOR), funded by the Ministry of Science and Culture of Lower Saxony and the Volkswagen Foundation, the research work focuses on cross-linked autonomous electric vehicles with fuel cells serving as range extenders. The aim of the project is the conception of digitized traffic systems that consist of cross-linked autonomous electric vehicles with range extender (RE). This contribution presents an approach for a model predictive power distribution to the energy sources battery and fuel cell considering their specific characteristics.

1. Introduction

The transdisciplinary joint project "ZuFOR" at Ostfalia University comprises three technical subprojects whose focuses are on innovative vehicle technologies in active safety systems for passenger cars, lightweight plastics structures in car manufacturing, and intelligent, digitized, cross-linked electric vehicles with Range Extender, respectively. The technical subprojects are complemented by an integrative subproject designed to facilitate and intensify knowledge transfer between project groups and business, research institutions as well as networks.

The aim of the subproject iREX 4.0 is the development of a flexibly scalable, predictive, electronic vehicle management (peVM) that comprises predictive algorithms for autonomous, energy-optimized operation. As regards operation strategy, there may be three potentials for energy-optimization when an electric vehicle is equipped with a fuel cell as range extender. The first potential lies in the optimization of the route guidance by means of an intelligent navigation algorithm that uses detailed map data comprising height profiles and data from V2X communication to determine the route in consideration of energy consumption and other constraints, e.g., available battery charge and hydrogen quantity or stations for recharging and filling up. The next potential is the optimization of the velocity profile to travel the selected route considering energy consumption as well as static constraints like speed limits and dynamic constraints like traffic lights. The third potential for optimization can be found in the two energy sources. The overall efficiency of the energy supply depends on the power distribution between fuel cell and battery in view of their unequal efficiency characteristics. As the future velocity profile is known, the required vehicle power can be determined in advance and the optimal power distribution can be specified from beginning to end of the journey. The present contribution will detail the approach of the iREX 4.0 subproject aiming at the third optimization potential to gain an energy-optimized power distribution at the level of the on-board power system.

IOP Conf. Series: Materials Science and Engineering **707** (2019) 012011 doi:10.1088/1757-899X/707/1/012011

2. Vehicle topology

The vehicle topology of this project (Figure 1) comprises an electric vehicle with one electric drive close to either of the wheels on the rear axle; every drive is mechanically linked to its wheel by a planetary gearbox and disposes of a bi-directional power electronics that will act as rectifier or converter and thus enable both the supply to the drive and the current feedback into the intermediate circuit (IC). The IC can be supplied with energy either by a uni-directional DC converter out of the fuel cell or by a bi-directional one out of the high-voltage battery. By means of the bi-directional DC converter, energy fed into the IC can be transported into the high-voltage on-board electric circuit as and when required. The low-voltage circuit with its own battery for provisioning low-power components is supported by the high-voltage circuit. A conventional hydraulic braking system will ensure safe braking if recuperative braking through the electric drives falls far short because of low speed or narrow machine limits. By means of individual steer-by-wire modules at each wheel of the front axle, it will be possible to carry out not only conventional steering maneuvers but also active toe-angle adjustments. For this purpose a central information processing unit is required that processes sensor- and V2X data and generates reference values for the actuator systems of the vehicle. During an autonomous drive, it will take over the driver's tasks and ensure safe lateral and longitudinal control.

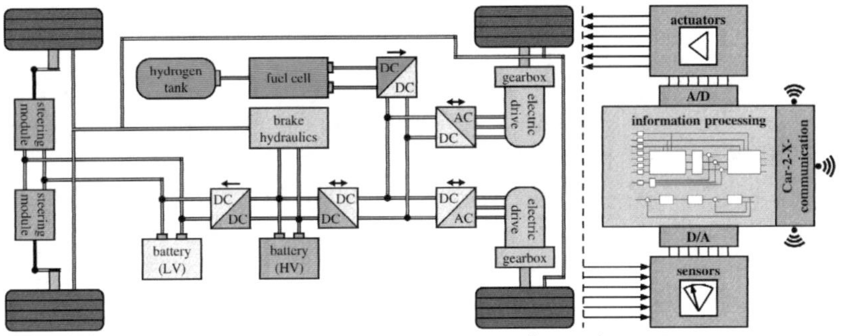

Figure 1. Focussed vehicle configuration [1].

3. Methodology

Increasing demands on and the complexity of vehicle systems make a clearly structured design methodology indispensable. The methodology employed is based on a mechatronic structuring of modules and hierarchies through a top-down procedure. The complex overall system is divided into intelligent, encapsulated subsystems, consisting of mechatronic components with defined interfaces, and then structured hierarchically. Figure 2 displays as an example for the focused vehicle topology (chapter 2) the resulting mechatronic structure of the research vehicle FREDY (cf. chapter 6.2 and [2]) comprising four hierarchical levels according to [3]: Mechatronic Function Modules (MFM), Mechatronic Function Groups (MFG), Autonomous Mechatronic Systems (AMS) and Cross-linked Mechatronic Systems (CMS). The MFM constitutes the lowest hierarchical level. It is a mechatronic system that cannot be divided further and consists of a mechanical supporting structure, sensor and actuator system, and information processing. This encapsulated module fulfils a defined functionality, maps the dynamical system behavior and represents kinematics, dynamics as well as the mechatronic functions. A mechanical or informational coupling of several MFMs yields MFGs having their own information processing that can in turn be combined to make up AMSs. Interlinking several AMSs to make up a CMS constitutes the top hierarchical level. A vehicle is an AMS that, through digitization and IoT, can be cross-linked with other vehicles to make up a CMS that is the basis for autonomous driving and nothing else than a cyber-physical system. Following the mechatronic structuring of the overall system, the mechatronic composition is used for developing the individual encapsulated modules in a bottom-up process and integrating them to constitute the higher-level overall system.

The mechatronic structure serves as the basis for deriving the structure of the hierarchical information processing (cf. Figure 3). At the core, there is the predictive electronic vehicle management (peVM) that represents the highest instance in the vehicle. It is part of a cyber-physical traffic system where the vehicle is linked with other traffic participants and the infrastructure via wireless communication (e.g., in ad-hoc networks according to IEEE 801.11p or mobile radio) and exchanges information with them. The peVM assumes the role of the human driver and plans route guidance on the basis of available information about driving states and environment conditions. Route guidance relies on detailed map data and has to take into account energy consumption, recharging infrastructure as well as actual traffic events to calculate the optimal route and project a velocity profile. It serves for deriving reference values for longitudinal and lateral control. These reference values are transmitted to the subordinated vehicle management, which provides the vehicle-dynamics control and the distribution of variables while comprising further chassis-assistance functions, and to the predictive energy management. The latter itself consists of energy assistance functions for predicting future on-board operation states and power requirements, the thermal energy management for controlling and optimizing heat streams, and the electric energy management with subordinated functions such as management of battery, fuel cell, electric drives or on-board system. These subordinated functions are used for monitoring and safeguarding as well as operating the respective aggregates on a local level. Moreover, electric energy management comprises the power distribution between battery and fuel cell for the purpose of increasing efficiency in the drivetrain.

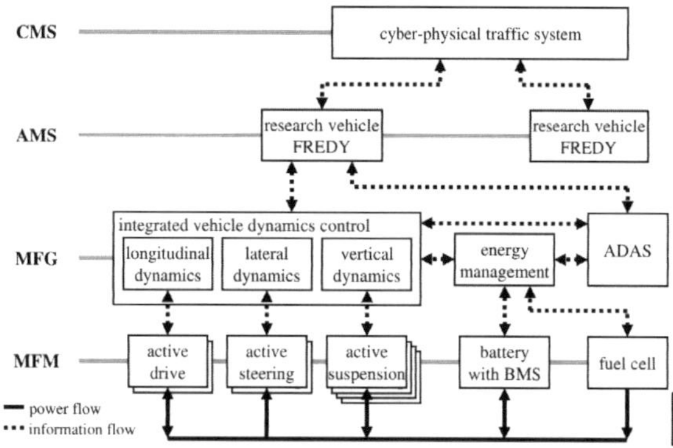

Figure 2. Mechatronic structuring of FREDY [4].

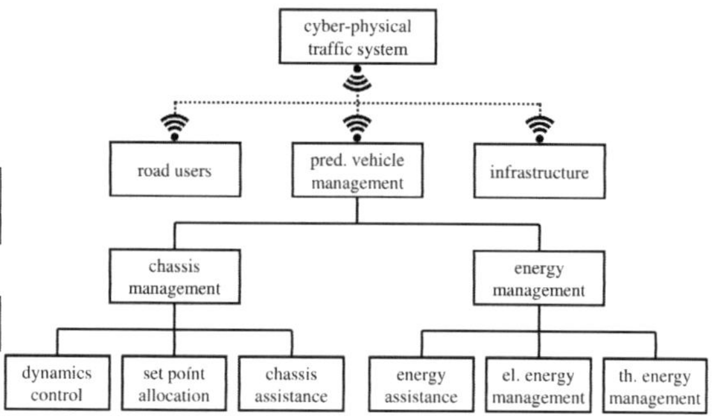

Figure 3. Hierarchical structure of the peVM within a cyber-physical traffic system [1].

Every function module with its associated information processing is designed according to the mechatronic development cycle (Figure 4). Theoretical investigation starts out from modelling which results in a physical resp. mathematical model, its behavior and validated parameters. This model is the basis for the model-based systems engineering of control strategies. This method is characterized by early validation and verification so that development time and costs can be reduced. It is done according to Rapid Control Prototyping (RCP) in a consistent, verification-oriented process with Model-in-the-Loop (MiL), Software-in-the-Loop (SiL) and Hardware-in-the-Loop (HiL) simulations.

A virtual test bench (Figure 5) is used for these MiL and SiL simulations. The algorithms of an Electronic Control Unit (ECU) can be tested automatically uesing a large database of virtual models and test cases. Results are visualized not only in diagrams but also in videos of the vehicle behavior so that the impact of changes to the algorithms can be comprehended directly.

A flexibly configurable HiL test bench (Figure 6) is designed for the purpose of a model-based, consistent design, validation and safeguarding. Its modular configuration enables treating various applications, from parameter identification and validation of simulation models to the testing, verification, and optimization of functions developed in the process. The test bench is based on the vehicle topology detailed in chapter 2. The test bench consists of a real-time module (RTM), a software module (SWM) and several modular test bench modules (TBM). The RTM performs the processing and digitization of the measurement signals from the TBM, computation of the control algorithms on a digital signal processor (DSP), setting reference values for the TBM actuators as well as communication with the TBM whereas the SWM comprises models of components that are not available, control algorithms, and the interface between user and test bench for visualizing the state and test bench control. The models and controllers are uploaded to the DSP via automated code generation. The TBM are made up of a fuel cell-, a battery-, a drive- as well as a load- and source-module, all of them operating in encapsulated mode and laid out for following application scenarios:

- **Fuel cell module:** Test of fuel cells for parameter identification.
- **Battery module:** Test and optimization of battery cells and -systems with defined climatic conditions, both via DC measurements and electrochemical impedance spectroscopy.
- **Drive module:** Test of el. drives and power electronics, trials and optimization of drive controls.
- **Load and source module:** This auxiliary module yields controllable supply voltages and supply currents as well as controllable load currents for simulating on-board components.

With suitable interfaces in a coupled operation of several test bench modules, many other application scenarios can be used:

- **Battery, fuel cell, and load/source modules:** Test of control concepts for power distribution between battery and fuel cell.
- **Battery, drive, and load/source modules:** Test of recuperation algorithms.
- **Fuel cell, battery, drive, and load/source modules:** Test of the drivetrain, involving fuel cell, battery, drive, and source as well as load that represent further components of the on-board system.

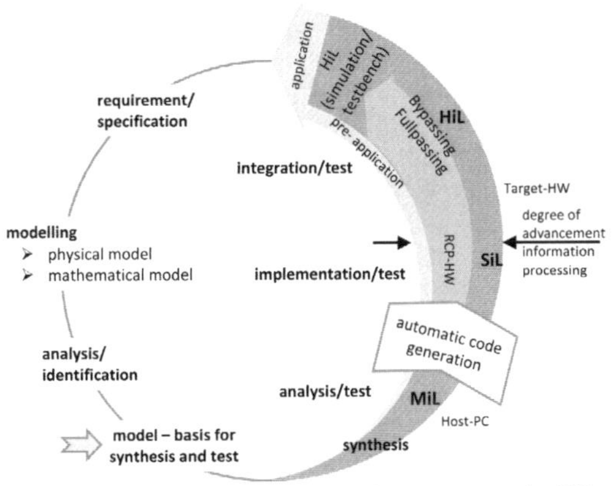

Figure 4. Mechatronic development cycle [2].

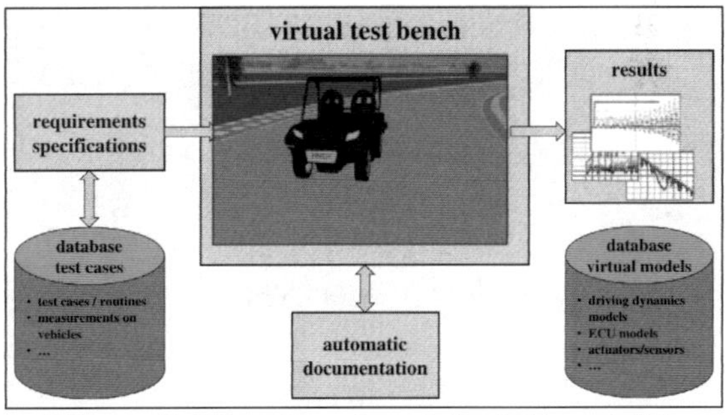

Figure 5. Virtual test bench for MiL and SiL simulations [4].

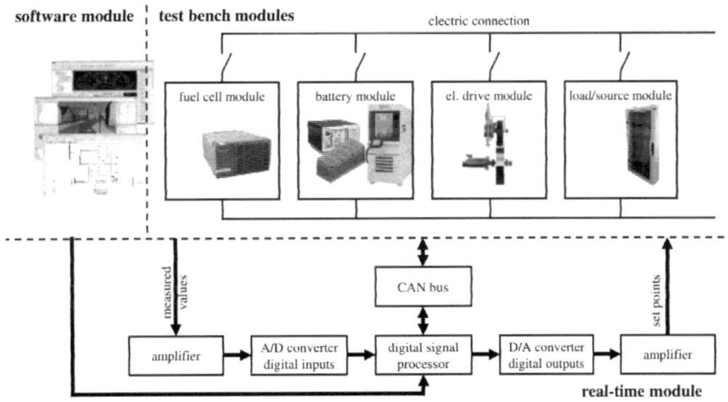

Figure 6. Modular structure of the flexible HiL test bench [1].

4. Modeling

This section shows the modeling of the lithium-ion battery and the fuel cell as the basis for the energy-optimized power distribution.

4.1. Battery

Lithium-ion cells are electrochemical energy storages, which can provide electric energy. The lithium-ion battery cell can be described with an equivalent circuit diagram according to [5]. This type of model is distinguished by a high accuracy, relatively easy ways to identify the parameters and a short computation time, which allows real-time applications. The cell behaves highly nonlinear and the terminal voltage v depends on the current i as well as on the actual state of charge SOC, the temperature T and the state of health SOH. The effect of the state of health is neglected in this first approach on modelling. The state of charge can be computed with the balance equation (1) with an initial state of charge SOC_0, the Coulombic efficiency η_C and the battery capacity C_n:

$$SOC = SOC_0 + \int \frac{\eta_C}{C_n} i \, \mathrm{dt} \tag{1}$$

The terminal voltage can be determined from the equivalent circuit diagram (Figure 7). The terminal voltage depends nonlinear on temperature and state of charge and consists of the open-circuit voltage OCV, ohmic losses on a serial resistance R_S and the voltages $v_{RC,i}$ on four RC elements to approximate the battery dynamics. For the terminal voltage follows

$$v = OCV - R_S i - v_{RC1} - v_{RC2} - v_{RC3} - v_{RC4} \tag{2}$$

Equation (2) can be used to determine the transfer function of the battery in the frequency domain, which shows the terminal voltage depending on the current:

$$G(j\omega) = \frac{v(j\omega)}{i(j\omega)} = R_S + \frac{R_1}{R_1 C_1 j\omega + 1} + \cdots + \frac{R_4}{R_4 C_4 j\omega + 1}. \tag{3}$$

Equation (3) shows that the current is the input and the terminal voltage the output of the battery model. The identification to quantify the parameters was performed both in time and frequency domain [6]. Figure 8 shows the OCV depending on the SOC as a result of measurements in time domain (Figure 8a) and the impedance behavior of the batteries at three exemplary SOCs (Figure 8b) as result of an Electrochemical Impedance Spectroscopy.

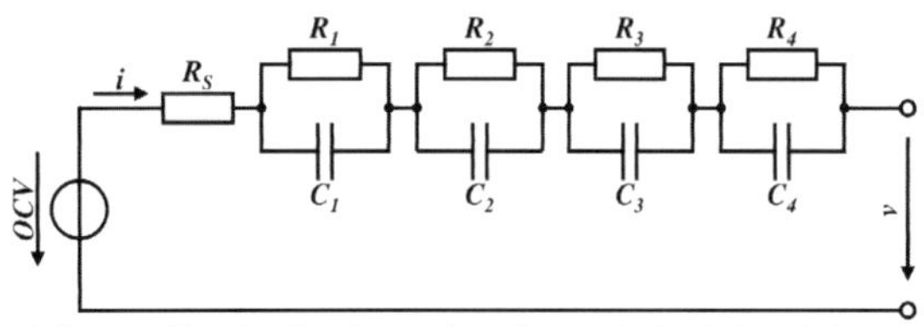

Figure 7. Equivalent circuit model of the li-ion battery [6].

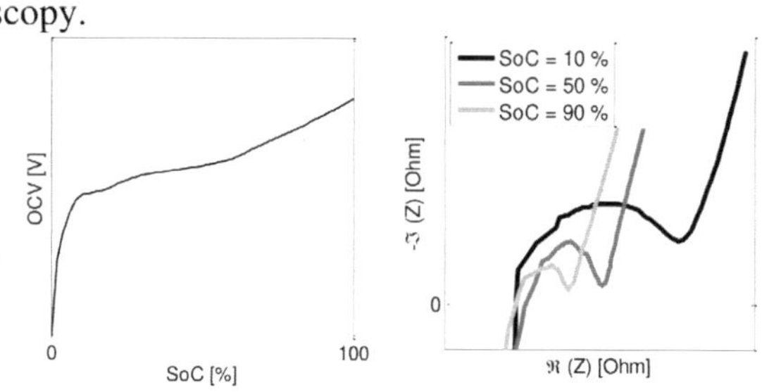

a) OCV diagram b) Nyquist diagram
Figure 8. Exemplary identification result of the battery parameters in time domain (a) and frequency domain (b) [6].

4.2. Fuel cell

Fuel cells are electrochemical energy sources, which can provide electric energy from hydrogen. A physics-based model of the stationary terminal voltage with an extension of electrical components to approximate the dynamic behavior is used to model the fuel cell (Figure 9), since this model offers a good compromise between accuracy and computational effort compared to 3D models and because it has a reference to the physics in contrast to empirical approaches.

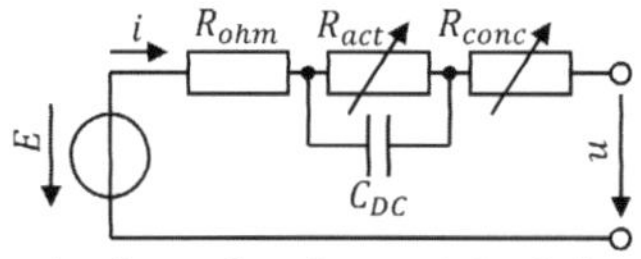

Figure 9. Equivalent circuit model of the fuel cell [7].

The polarization voltage (stationary terminal voltage) is modelled using the Nernst equation and relevant overvoltages. The Nernst equation

$$E = E^0 + \frac{RT}{2F} \ln\left(\frac{p_{H_2} \cdot p_{O_2}^{0,5}}{p_{H_2O}}\right) \tag{4}$$

describes the potential difference E between the electrodes as a function of the standard electrode potential E^0, the fuel cell temperature T and the pressures p_{H_2}, p_{O_2} and p_{H_2O}. The pressures are controlled by the fuel cell system and therefore considered constant in this work. The general gas constant R and Faradays constant F are also used in that equation. The terminal voltage of the fuel cell

$$u = E - u_{act} - u_{ohm} - u_{conc} \tag{5}$$

results from Nernst voltage E minus activation overvoltage u_{act}, ohmic voltage drop u_{ohm} and concentration overvoltage u_{conc}. The activation overvoltage can be described using the Tafel equation

$$u_{act} = \frac{RT}{2\alpha F} \ln\left(\frac{i}{i_0}\right), \tag{6}$$

that uses the charge transfer coefficient α, the current i and the constant i_0. The ohmic voltage drop

$$u_{ohm} = R_{ohm} i \tag{7}$$

represents losses on the internal resistance R_{ohm}. The concentration overvoltage

$$u_{conc} = -\frac{RT}{2F} \ln\left(1 - \frac{i}{i_l}\right) \tag{8}$$

has a significant influence on the terminal voltage only at high currents close to the limiting current i_l due to decreasing concentrations of the working gases as well as time-limited transport processes within the fuel cell. The electrical equivalent circuit diagram (Figure 9) shows the extension of the stationary model with a capacity C_{DC}, which represents the double-layer capacity of the fuel cell. The extended model is a PT1 system

$$u_{act} + R_{act} C_{DC} \dot{u}_{act} = R_{act} i \tag{9}$$

with the gain R_{act} and the time constant $\tau = R_{act} C_{DC}$ for an approximation of the dynamics. The parameters of the fuel cell are identified by measurements in time and frequency domain gained using the flexible HiL test bench [7] analogous to the battery parameters. Figure 10 shows an exemplary identification result in the time domain.

a) Comparison of measured and simulated polarization voltage

b) Measured stack temperature

Figure 10. Exemplary identification result of the fuel cell parameters in time domain [7].

It shows a comparison of measured and simulated polarization voltage of the fuel cell (Figure 10a) and the fuel cell temperature during the measurement (Figure 10b). The temperature increases with increasing load and duration starting from the equilibrium and decreases again with decreasing load until the equilibrium is reached again. This behavior is reflected in the polarization voltage, which also assumes higher voltages for higher temperatures. The voltage reaches the equilibrium after the fuel cell returned to its temperature equilibrium. The comparison of measured and simulated polarization voltage shows the high accuracy of the model with the identified parameters. The following section deals with the model predictive power distribution (MPPD) based on the battery and fuel cell models.

5. Model predictive power distribution

5.1. Power prediction

Energy and time optimized velocity profiles (cf. [8]) are the reference values for the longitudinal control of autonomous vehicles. Such an optimized velocity profile is used for calculating a power profile of the vehicle for the entire route, so that the power can be distributed between fuel cell and battery in a next step. At first, assuming constant acceleration, the velocity profile between two nodes is transformed in such a way that the velocity no longer depends on the distance but on time. For prediction (cf. Figure 11), an inverse vehicle-dynamics model is used to determine the tire circumference forces in longitudinal direction $F_{x,i}$ from the predicted velocity profile v_{pred}. By means of inverse tire models, the resulting curves mapping the longitudinal forces are then converted into wheel-angle-speed- and wheel-torque curves, $\omega_{x,i}$ and $M_{x,i}$; an inverse drivetrain model is than used for a prediction of the mechanical and finally electrical driving power profile P_{pred}.

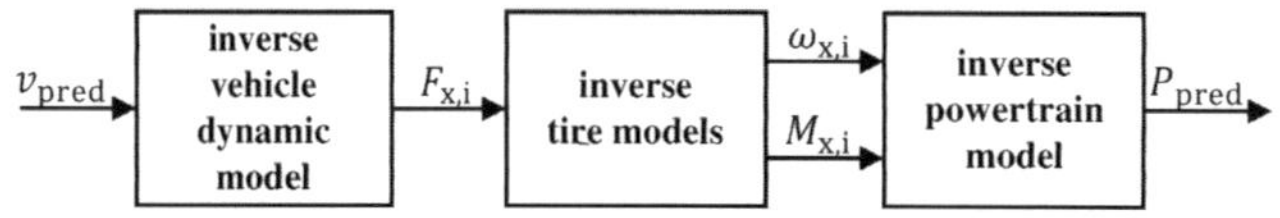

Figure 11. Structure of the power prediction.

Taking a drive from Ostfalia University in Wolfenbuettel to Ostfalia in Wolfsburg as an example, the map displays the predicted velocity profile (Figure 12a) and the predicted power profile calculated from the latter (Figure 12b). It is obvious that during acceleration both positive power peaks (increase in speed) and negative ones (recuperation) occur whereas in phases of constant velocity the power is also constant. In the following, the predicted power profile will be the ref. value for MPPD.

a) Predicted velocity profile b) Predicted power profile

Figure 12. Exemplary result of power prediction (b), based on the predicted velocity profile (a) for a drive from Ostfalia in Wolfenbuettel to Ostfalia in Wolfsburg.

5.2. Control algorithm

The optimization problem of the MPC concerns the distribution of the predicted power between battery and fuel cell to minimize energy losses on the whole route until the travel destination is reached. For solving the optimization problem, a model predictive control (MPC) is laid out, because it takes into account the future system behavior and improves it by selecting an optimal sequence of ref. values. The first approach to a MPPD simplifies the problem in that battery and fuel cell voltages are constant and models are linearized. The relevant system states

$$\underline{x}_k^T = [SOC_k \quad m_{H_2,k}] \tag{10}$$

describe the state of charge of the battery SOC and the filling level of the hydrogen tank m_{H_2}. The vector of manipulated variables

$$\underline{u}_k^T = [i_{bat,k} \quad i_{fc,k}] \tag{11}$$

contains the currents of the two energy sources that have an effect on the states. Future behavior of the system states can be calculated using the linear state equation

$$\underline{x}_{k+1} = \underline{\underline{A}}\,\underline{x}_k + \underline{\underline{B}}\,\underline{u}_k. \tag{12}$$

This calculation will be performed in the following time steps from $k+1$ to the end of the prediction horizon $k+N_p$, with the results being summed up in the vector of future states

$$\underline{\tilde{x}}_k^T = [\underline{x}_{k+1} \quad \underline{x}_{k+2} \quad \cdots \quad \underline{x}_{k+N_p}]. \tag{13}$$

From the current time k to the end of the control horizon N_u, which is equated with the prediction horizon N_p, the future variables are accumulated in the vector of future manipulated variables

$$\underline{\tilde{u}}_k^T = [\underline{u}_k \quad \underline{u}_{k+1} \quad \cdots \quad \underline{u}_{k+N_u-1}] = [\underline{u}_k \quad \underline{u}_{k+1} \quad \cdots \quad \underline{u}_{k+N_p-1}]. \tag{14}$$

From the future manipulated variables, the future overall current

$$\underline{\tilde{i}}_k^T = [i_{sum,k+1} \quad i_{sum,k+2} \quad \cdots \quad i_{sum,k+N_p}] \tag{15}$$

can be calculated based on the general correlation

$$i_{sum} = i_{bat} + i_{fc} \tag{16}$$

that is used with all time steps. This yields the matrix

$$\tilde{\underline{i}}_k = \begin{bmatrix} 1 & 1 & \cdots & 0 & 0 \\ \vdots & \vdots & \ddots & \vdots & \vdots \\ 0 & 0 & \cdots & 1 & 1 \end{bmatrix} \tilde{\underline{u}}_k. \tag{17}$$

Under the above-mentioned assumption of the constant voltages, the future overall power

$$\tilde{\underline{P}}_k = \begin{bmatrix} u_{bat} & u_{fc} & \cdots & 0 & 0 \\ \vdots & \vdots & \ddots & \vdots & \vdots \\ 0 & 0 & \cdots & u_{bat} & u_{fc} \end{bmatrix} \tilde{\underline{u}}_k = \underline{\underline{K}}\tilde{\underline{u}}_k \tag{18}$$

of the energy sources can be calculated. The MPC has the aim of distributing the predicted power

$$\tilde{\underline{P}}^T_{pred,k} = \begin{bmatrix} P_{pred,k+1} & P_{pred,k+2} & \cdots & P_{pred,k+N_p} \end{bmatrix} \tag{19}$$

between battery and fuel cell over the entire drive while minimizing the energy losses

$$E_{L,k} = \underline{u}_k^T \underline{\underline{R}}\,\underline{u}_k \tag{20}$$

that accrue until the end of the drive. For reasons of simplification, the losses are assumed to occur only at the internal resistances of the energy sources, comprised in $\underline{\underline{R}}$. The composed cost function

$$J_k(\tilde{\underline{u}}_k) = g_P\left(\tilde{\underline{P}}_{pred,k} - \tilde{\underline{P}}_k\right)^T\left(\tilde{\underline{P}}_{pred,k} - \tilde{\underline{P}}_k\right) + g_E E_{L,k} = g_P\left(\tilde{\underline{P}}_{pred,k} - \underline{\underline{K}}\tilde{\underline{u}}_k\right)^T\left(\tilde{\underline{P}}_{pred,k} - \underline{\underline{K}}\tilde{\underline{u}}_k\right) + g_E \tilde{\underline{u}}_k^T \underline{\underline{R}}\tilde{\underline{u}}_k \tag{21}$$

is the basis for optimization and cosinders the mean square deviation between the reference and actual values as well as the energy losses. Through a transformation and by neglecting components that cannot be impacted by $\tilde{\underline{u}}_k$, the cost function can be converted into the quadratic optimization problem

$$J_k(\tilde{\underline{u}}_k) = \tilde{\underline{u}}_k^T(g_P \underline{\underline{K}}^T \underline{\underline{K}} + g_E \underline{\underline{R}})\tilde{\underline{u}}_k - 2 g_P \tilde{\underline{P}}^T_{pred,k} \underline{\underline{K}}\tilde{\underline{u}}_k. \tag{22}$$

The optimization has the aim of minimizing the cost function

$$\min_{\tilde{\underline{u}}} J_k(\tilde{\underline{u}}_k) = \min_{\tilde{\underline{u}}}\left(\tilde{\underline{u}}_k^T(g_P \underline{\underline{K}}^T \underline{\underline{K}} + g_E \underline{\underline{R}})\tilde{\underline{u}}_k - 2 g_P \tilde{\underline{P}}^T_{pred,k} \underline{\underline{K}}\tilde{\underline{u}}_k\right) \tag{23}$$

by means of an optimal sequence of variables $\tilde{\underline{u}}_k^*$ and taking into account the inequality constraints

$$\underline{\underline{N}}\tilde{\underline{u}}_k \geq \underline{g}. \tag{24}$$

The inequality constraints $\underline{g}$ comprise the operating limits of battery and fuel cell current as well as the minimum and maximum admissible states (battery SOC and hydrogen in the tank). One drawback of MPC with constraints is the fact that the optimization problem can no longer be solved analytically, thus requiring a numerical solution. There are defined algorithms that can be used for solving quadratic optimization problems. Nonlinear behavior or cost functions that do not refer to a quadratic optimization problem can be solved by means of dynamic programming according to Bellman.

5.3. Results

Figure 13 displays an exemplary first scenario of the MPPD. The predicted power curve that was set as reference value is realized as the sum of fuel cell and battery power (a). The power is divided to the battery (b) and the fuel cell (c) while only the battery is able to store negative power from recuperation. Due to the load of the driving task, the battery SOC (d) as well as the hydrogen stored in the tank (e) decrease. The energy losses reach an overall optimum according the ref. power profile.

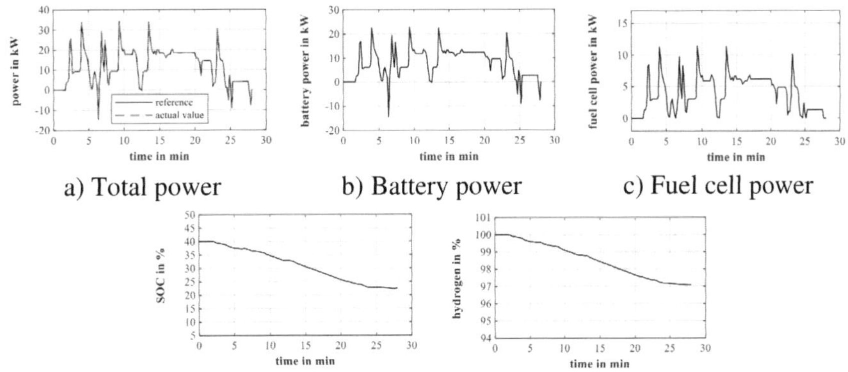

a) Total power b) Battery power c) Fuel cell power

d) State of charge e) Hydrogen tank

Figure 13. Exemplary result of the model predictive power distribution.

Figure 14 displays another exemplary result of the MPPD with reaching a state limit. The same predicted power curve (a) was set as reference value in a second scenario and is again realized as the sum of fuel-cell- and battery power. The overall battery power (c) of the second scenario is lower than in the first one so that in reverse the fuel cell power (d) is higher in the second scenario. This distribution avoids a violation of the lower battery SOC limit of 10 % (e) so that there will be no deep discharging that may damage the battery in the long run. In order to achieve the overall power and to meet this constraint, higher power is extracted from the fuel cell that disposes of enough fuel (f). The energy losses (b) reach an optimum only according the ref. power profile and the constraints. Indeed, this optimum is not the overall one that can only be reached with a higher initial SOC (cf. scenario 1).

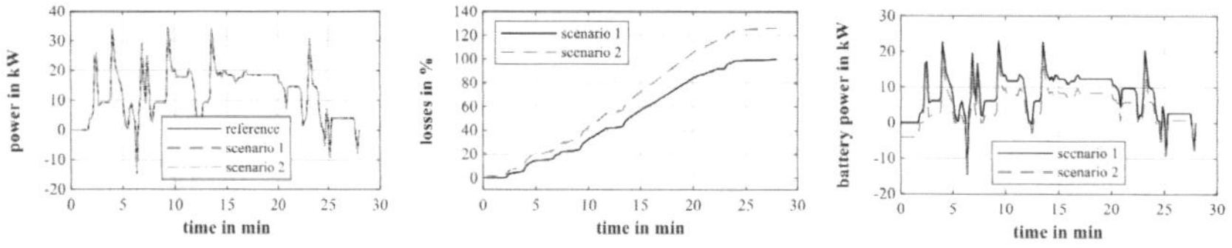

a) Total power b) Losses c) Battery power

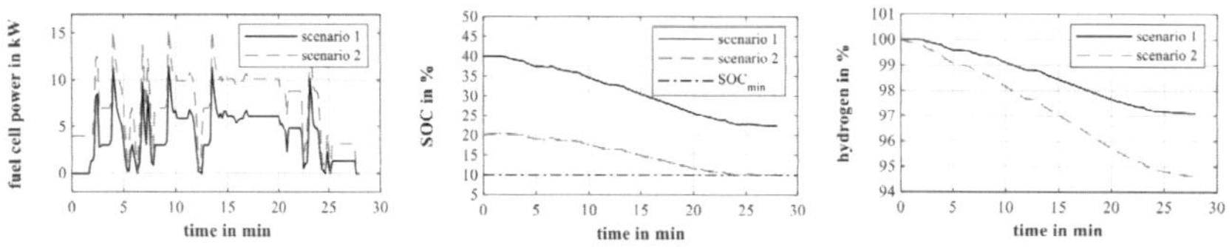

d) Fuel cell power e) State of charge f) Hydrogen tank

Figure 14. Exemplary results of the model predictive power distribution incl. reaching a constraint.

6. Realization

6.1. Flexible HiL test bench

Figure 15 displays the realized HiL test bench (cf. chapter 3) with exception of the drive module. At the present stage of development, the individual modules are functioning so that the test bench can be used for tests on every single component. The HiL test bench was used to perform the parameter and system identification as well as the validation and optimization of the battery and fuel cell models. Further work on the project will concentrate on the integration of the components with the aim of realizing the combined applications. Therefore, especially innovative power electronics are under to perform investigations on the MPPD under real-time conditions.

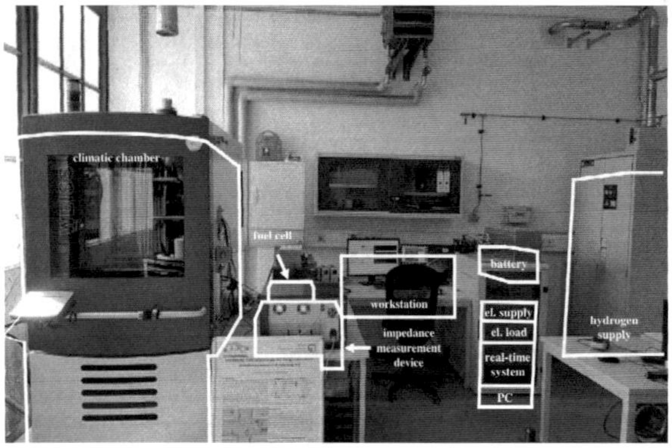

Figure 15. Realization of the flexible HiL test bench [1].

Figure 16. Current state of completion of the research vehicle FREDY.

6.2. FREDY

The Function Carrier for Regenerative Electromobility and Vehicle Dynamics, FREDY, serves the purpose of validating and optimizing vehicle functions in reality when these have been sufficiently examined and optimized by means of simulations and test bench trials. It is an electric vehicle that conforms to the concept displayed in Figure 1, except for the fuel cell, which will be mounted later. FREDY (Figure 16) disposes of two electric drives close to the wheels that get their power from two lithium-ion battery packs via bi-directional power electronics. A dSPACE MABX II is the central information-processing unit and performs all measurement and control tasks. Future work will include realization, validation and optimization of the MPPD after the integration of the fuel cell to FREDY.

7. Summary and outlook

The present contribution describes the underlying problem and range of tasks in the iREX 4.0 project where intelligent algorithms for an energy-optimal drive of electric vehicles with fuel cells are to be developed. A consistent, model-based design methodology for mechatronic systems was introduced and applied on the iREX 4.0 project. Moreover, a flexibly configurable HiL test bench was presented that will be employed at every project stage, from parameter identification of simulation models to validation and optimization of developed algorithms in a real-time environment. It was a given a brief overview on modeling a lithium-ion battery and a PEM fuel cell in order to achieve a high accuracy combined with a low computational effort. A power profile is predicted based on an optimized velocity profile as reference value for the autonomous driver. This power profile is used by a model predictive control that distributes the overall power to the battery and the fuel cell whilst maintaning the operation limits in order to achieve an optimum concerning the overall energy losses.

Future activities will include extending the MPPD to consider also the non-linear behavior of battery and fuel cell. Moreover, parallel computation of the MPPD using different timescales will be focused; thus, computation for a short time horizon from the actual position will accordingly be performed more detailed with a shorter sample time than a longer time horizon to the end of the journey because the computation for the long period of time has greater uncertainties. Furthermore, we will proceed to realize the MPPD also with the flexibly configurable HiL test bench and the FREDY research carrier.

8. Acknowledgement

The present contribution, elaborated in the context of the project „ZuFOR“, was funded by the Ministry of Science and Culture of Lower Saxony and the Volkswagen Foundation. Grant number: VWZN3236. Responsibility for the content of this paper lies with the authors.

Niedersächsisches Ministerium für Wissenschaft und Kultur VolkswagenStiftung

9. References

[1] Scherler S et al. 2018, *Model-based design of a multi-functional HiL test bench for investigations on a Range Extended Vehicle*, 4th IEEE Int. Symp. on Systems Engineering.

[2] Liu-Henke X et al. 2015, *FreDy – an electric vehicle with intelligent chassis control systems*, 10th IEEE Int. Conference on Ecological Vehicles and Renewable Energies.

[3] Liu-Henke X 2004, *Mechatronische Entwicklung der aktiven Feder-/Neigetechnik für das Schienenfahrzeug RailCab*, VDI-Fortschritt-Berichte, Reihe 12, Nr. 589.

[4] Liu-Henke X et al. 2016, *Holistic development of a fullactive electric vehicle by means of a model-based systems engineering,* 2nd IEEE Int. Symp. on Systems Engineering.

[5] Tsang K M, Chan W L, Wong Y K and Sun L 2010, *Lithium-ion Battery Models for Computer Simulation*, IEEE International Conference on Automation and Logistics.

[6] Quantmeyer F and Liu-Henke X 2014, *Modeling the Electrical Behavior of Lithium-Ion Batteries for Electric Vehicles*, Solid State Phenomena Vol. 214, pp. 40 – 47.

[7] Scherler S, Liu-Henke X and Henke M 2019, *Modellbildung und Systemidentifikation einer PEM-Brennstoffzelle*, Workshop der ASIM/GI-Fachgruppen.

[8] Fritsch M and Liu-Henke X 2017, *Optimization of energy consumption by using an intelligent assistance system for an electric vehicle*, 12th IEEE Int. Conference on Ecological Vehicles and Renewable Energies.

IOP Conf. Series: Materials Science and Engineering **707** (2019) 012012 doi:10.1088/1757-899X/707/1/012012

Obstacle detection using ultrasonic sensor for a mobile robot

Joseph Azeta[1, 2], Christian Bolu[1], Daniel Hinvi[1] and Abiodun A Abioye[1]

[1]Department of Mechanical Engineering, Covenant University, Ota, Nigeria.

Abstract. There have been a number of successful attempts in designing obstacle avoiding robots. These works differ by selection of sensors, path mapping process and the algorithms applied to set the operational parameters. In this paper we present a los cost ultrasonic distance sensor for obstacle avoidance for mobile robot navigation. The system is implemented using microcontroller Arduino Uno, a Wi-Fi module, an Arduino motor shield driver which controls the robot through the geared dc motors. The system showed good performance under various lighting conditions. Experimental results with varied positions of obstacle show the flexibility of the robot to avoid it and have shown a decent performance in our laboratory. The robot is additionally ready to acknowledge victims before it, the sensing element system is extremely low-cost as a result of it solely uses one distance sensing element.

1. Introduction

There have been a number of successful attempts in designing obstacle avoiding robots [1, 2, 3]. These works differ by selection of sensors, path mapping process and the algorithms applied to set the operational parameters. There have been numerous projects in this area using laser scanner, an infrared sensor, GPS and multiple sensors to accomplish obstacle detection and avoidance [4]. Researchers are persistently trying to find more precise ways to develop autonomous robot or vehicle movement technology. In obstacle detection, the selection of sensor is vital for the required application of the robot, otherwise, it might fail to operate even though all hardware and software are working properly. For example, a robot with optical sensors in a room with glass walls might create more collisions than avoidance. Hence sensors should be selected in accordance with the characteristics of the obstacles. 240° laser scanner was used as a sensor to build a robot based on Small Mobile Robot (SMR) platform. The robot generates a collision-free path from a grid map using wavefront algorithm. Other researchers applied low cost sensor network for avoiding obstacle using RGB and Infrared camera and ultrasonic sensors which can detect different obstacles [5] also obstacle warning and avoidance system is an important technology in several applications including manned and unmanned aircraft [6]. [7] Present tree trunk detection and mobile robot localization using camera and ultrasonic sensor to get the location data of the trunk. Proposed road detection and tracking system for mobile robots which achieved a good recognition rate [8].

2. Related Works

[9] Presented a single-beam mechanically-scanning profiling sonar to detect obstacles under water. The profiling sonar has the ability to produce a cone-shaped beam which is ideal for detecting near-surface obstacles. One of the objectives of their work was to investigate the suitability of using sonar near the water-air boundary for which the study found promising results. There were other works using multiple sensors to make the robot more accustomed to its surroundings by employing both range and appearance-based obstacle detection [10, 11]. Their obstacle detection also includes a combination of global and local avoidance. They fused the strengths of an image and an ultrasonic sensor to detect objects and measure its size. Detection of the object was carried out by the ultrasonic

sensor and its measurement required the help of a camera. The code was designed to receive the distance to object, its height and width. Surveillance robots are specially designed for mining accidents, earthquake disasters, and hostage situations etc. Many groups of researchers have developed surveillance robots with the ability to conduct emergency search and rescue work. [12] presented 'Robot-Assisted Emergency Search and Rescue System with a Wireless Sensor Network. [13] presented 'Mobile Rescue Robot for Human Body Detection in Rescue Operation of Disaster'. [14] presented 'Rescue Robotics Using Artificial Intelligence', [15] presented 'Design and Implementation of e-Surveillance Robot for Video Monitoring and Living Body Detection' and [16] presented 'Surveillance Robot Using Arduino Microcontroller, Android APIs and the Internet'. All of those researchers are based on gas sensor, LDR & metal detector to detect a bomb, PIR to sense human movement and to detect human alive or dead, LM35 to detect temperature, Distance sensors, Compass & communicating over Zig-Bee, RF & internet. By detecting human body temperature wirelessly, the system can identify whether the human is alive or dead. The robot also has a long-range First-person view (FPV) video monitoring system.

3. Implementations

The system is implemented using microcontroller Arduino Uno, a Wi-Fi module, an Arduino motor shield driver which controls the robot through the geared dc motors. The Arduino is powered by a 7.4V, 2400mAh lithium polymer battery, which sends sizeable current to the dc motor for its movement and also powers the Wi-Fi module for visual transmission and recording of data which also send current to the servo motors for tilting the robot camera module for optimum visual.

3.1. Arduino UNO R3 Microcontrolller

The Arduino projects provide an integrated development environment (IDE) supported process, and programming is completely employing a language supported wiring that is extremely almost like C++. The Arduino Uno is the motherboard for which this robot depends on utilizing the ATmega328. It has fourteen advanced pins for info and yield. It is fueled by a DC supply or an AC to DC gadget. Inside the Arduino, there is a microcontroller that is a PC circuit having a memory, processor, and programmable info yield pins as shown in figure 1.

Figure 1. Arduino UNO R3

3.2. Engine Control Module (Motor Shield)

The Arduino Motor shield relies on the L298 (datasheet), that could be a dual full-bridge driver designed to drive inductive loads like relays, solenoids, DC and stepping motors. It allows you to drive two DC motors along with your Arduino board, controlling the speed and direction of every one independently. This shield has separate channels and uses four of the Arduino pins to settle on the rotation direction and varying the speed, quick break or sense the current flowing through the motor. In total there are eight pins in use on this shield. You will be able to use every channel on an individual basis to drive 2DC motors or more or mix them to drive one bipolar stepper motor as shown in figure 2.

Figure 2. PWR Power and Motor Shield

3.3. HC-SR04 Ultrasonic sensor
The HC-SR04 ultrasonic sensing element uses echo sounder to work out the gap to an object like dolphins do. It offers wonderful non-contact detection with high accuracy associated stable readings in an easy-to-use package from two cm to four hundred cm. Its operation isn't restricted by daylight or black material. It comes with a transmitter and recipient module shown in figure 3.

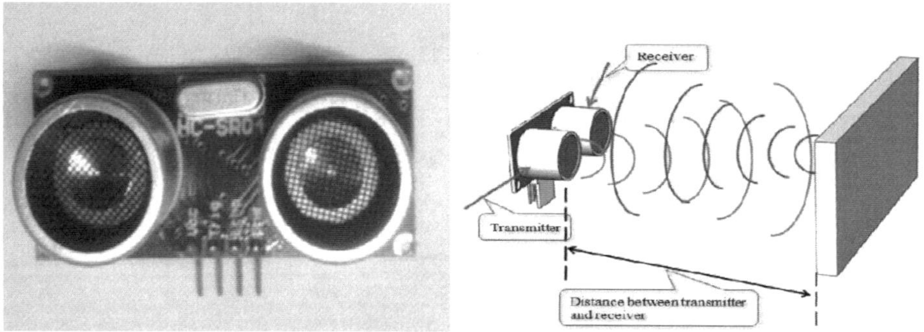

Figure 3. Ultrasonic Sensor

3.4. Working Principle
As shown in figure 3, the ultrasonic sensor radiates the short and high-recurrence flag. These spread inside the air at the rate of sound. The ultrasonic sensor comprises a multivibrator, attached to the base. The multivibrator is a mix of a resonator and vibrator. The resonator conveys imperceptible wave created by the vibration. The ultrasonic sensor truly comprises of two sections; the anode that creates a 40 kHz wave and consequently the indicator recognizes 40 KHz wave and sends an electrical flag back to the microcontroller. The ultrasonic sensor enables the robot to pretty much observe and recognize a protest, maintain a strategic distance from impediments.

3.5. Self-Protection
The collision avoidance is authorized utilizing an ultrasonic sensor; that transmits an unintelligible heartbeat each 0.3s to discover any obstruction inside the change of 100cm. considering the broadness of the system is 30cm; the indicator is guided by a servo at skillet points of - 30°; - 15°; 0°; +15°; + 30° to discover any hindrance close the trail of the robot; wherever 0° speaks to the bearing of straight movement ahead. It's accepted that there aren't any overhanging deterrents or sharp obstructions like metal bars. The ultrasonic sensor beneath the base of the system sporadically screens the space to the

base. The rough separation to the base is 4cm; though the component is on the base. Any positive deviation from this value is dealt with as a hoist up from the base and produces an alarm.

3.6. Power Supply

Powered supply via a 7.4V 2400mAh lithium polymer battery. Lipo is a reversible battery of lithium-ion technology employing a compound solution rather than a liquid one. High physical phenomenon solid polymers form this kind of solution. These lipo batteries give the next specific energy than alternative lithium-battery sorts.

4. Result

Table 1. Avoidance Accuracy of Static Obstacle

Environment	Type of obstacle	Detected	Avoided	Accuracy
Well-lit	Single solid obstacle	Yes	Yes	100%
Dimly-lit	Single solid obstacle	Yes	Yes	100%
Well-lit	Uniform shaped surface	Yes	Yes	100%
Dimly-lit	Uniform shaped surface	Yes	No	40%
Well-lit	Double solid obstacles	Yes	Yes	75%
Dimly-lit	Double solid obstacles	Yes	Yes	70%
			Average	80.83%

Accuracy = (number of successful avoidances)/(number of test cases)

Accuracy Range	
Extremely Sharp	80-100%
Slightly sharp	60-79%
Sharp	50-59%
Not sharp	0-49%

Table 1 shows the precision of the system when tested against the fluctuated environment, variety of obstacle, and in this way the things once the object is recognized it stayed away from it.
Experimental results with varied positions of obstacle show the flexibility of the robot to avoid it and have shown a decent performance in our laboratory. The robot is additionally ready to acknowledge victims before it, the sensing element system is extremely low-cost, as a result, it solely uses one distance sensing element.

5. Conclusion

Obstacle avoidance capability needs to be considered when designing mobile robots for different applications. The low cost ultrasonic sensor for mobile robot is aim to design and implement a helpful tool that improves the ability of mobile robot to avoid obstacle successfully. A series of test were done to check the reliability of the system. In our experiment the ultrasonic distance sensing element was accustomed to offer a large field of detection. Which can be implemented on mobile robots both remotely controlled and also on autonomous mode, once in the autonomous mode, the initial loading of the code needs no user intervention throughout its operation. When it is placed in an unknown setting with obstacles, it runs while avoiding all obstacles with significant accuracy. Result demonstrated high accuracy of the ultrasonic sensor to avoid obstacle

6. Acknowledgment

The Authors are grateful to Covenant University management for publication support.

7. References

[1] M. A. Ahasan, S. A. Hossain, A. U. Siddiquee & Rahman 2012 Obstacles Invariant Navigation of An Autonomous Robot based on GPS. Khulna University.

[2] Kam, M., Zhu, X. & Kalata, P 1997 Sensor Fusion for Mobile Robot Navigation. Proceedings of the IEEE 85(1) 108-119.

[3] Azeta Joseph, Bolu Christian, Abioye A. Abiodun and Festus Oyawale 2018 A review on humanoid robotics in healthcare. *MATEC Web Conf.*, 153 02004.

[4] C. A Ryther, & O. B. Madsen 2009 Obstacle Detection and Avoidance for Mobile Robots. The Technical University of Denmark.

[5] Pu, J., Jiang, Y., Xie, X., Chen, X., Liu, M. & Xu, S 2018 Low cost sensor network for obstacle avoidance in share-controlled smart wheelchairs under daily scenarios. *Microelectronics Reliability*, 83, 180-186.

[6] Ramasamy, S., Sabatini, R., Gardi, A. & Liu, J 2016 LIDAR obstacle warning and avoidance system for unmanned aerial vehicle sense-and-avoid. *Aerospace Science and Technology*, 55, 344-358.

[7] Chen, X., Wang, S., Zhang, B. & Luo, L 2018 Multi-feature fusion tree trunk detection and orchard mobile robot localization using camera/ultrasonic sensors. *Computers and Electronics in Agriculture* 147, 91-108.

[8] Zhang, H., Hernandez, D., Su, Z., & Su, B 2018 A low cost vision-based road-following system for mobile robots. *Applied Sciences*, 8(9).

[9] Heidarsson, H. K., & Sukhatme, G. S 2011 Obstacle Detection and Avoidance for an Autonomous Surface Vehicle using a Profiling Sonar. *IEEE International Conference on Robotics and Automation*. Shanghai.

[10] Shahdib, F., Ullah, M. W., Hasan, M. K., & Mahmud, H 2013 Obstacle Detection and Object Size Measurement for Autonomous Mobile Robot using Sensor. *International Journal of Computer Applications* 66(9) 28-33.

[11] K. W. Gray 2000 Obstacle Detection and Avoidance for an Autonomous Farm Tractor. Utah State University.

[12] Ko, A. & Henry, Y. K. L 2009 'Robot-Assisted Emergency Search and Rescue System with a Wireless Sensor Network'. *International Journal of Advanced Science and Technology*, 69-78.

[13] Bhondve, T. B., Satyanarayan, R. & Mukhedkar, M 2014 'Mobile Rescue Robot for Human Body Detection in Rescue Operation of Disaster', *International Journal of Advanced Research in Electrical, Electronics and Instrumentation Engineering* 3(6) 9876-9882.

[14] Seethai, A., Periasamy, A. & Muruganand, S 2013 'Rescue Robotics Using Artificial Intelligence'. *International Journal of Advanced Research in Computer and Communication Engineering* 2(12) 4589-4593.

[15] Dixit, K. S. & Dhayagonde, S. B 2014 'Design and Implementation of e-Surveillance Robot for Video Monitoring and Living Body Detection'. *International Journal of Scientific and Research Publications* 4(4) 1-3.

[16] Kulkarni, C., Suhas, G., Suresh, P. G., Krishna, C. & Antony, J 2014 'Surveillance Robot Using Arduino Microcontroller, Android APIs and the Internet', *First International Conference on Systems Informatics, Modelling and Simulation, IEEE Computer Society Washington* 107-111.

IOP Conf. Series: Materials Science and Engineering **707** (2019) 012013 doi:10.1088/1757-899X/707/1/012013

Finding verifying vertices of a coefficients polytope of characteristic polynomial for analyzing a robust oscillability of a control system with interval parameters

I V Khozhaev[1], **S A Gayvoronskiy**[1] **and T A Ezangina**[1]

[1] School of Computer Science and Robotics, National Research Tomsk Polytechnic University, Tomsk, 634034, Russia

Abstract. The paper is dedicated to extending a root locus theory to systems with interval parameters, which are described with a characteristic polynomial with interval coefficients. On a basis of angular equation of a root locus a set of interval inequalities, including departure angles of root locus edge branches, determining oscillability degree of a control system, was derived. Solving the system of inequalities for low-order control systems resulted in a set of vertices of a characteristic polynomial coefficients polytope, projections of which on a complex plane determine an oscillability of such systems. Examples of application are provided.

1. Introduction

Designing a control system with interval parameters consists in solving two types of problems: problem of synthesis and problem of analysis. Synthesizing an interval control system (ICS) is usually dedicated to providing a desired control quality despite parametric variations within certain intervals [1] – [10]. Analyzing an ICS control quality is dedicated to estimating robust control quality – a set of the worst values of quality indices, which can be reached during parametric variations [1] –[10]. A common approach to analyzing a robust control quality is a root approach, which is simple and illustrative [4].

Root approach is based on analysis and synthesis of a characteristic polynomial of a control system. For ICS let us introduce a term interval characteristic polynomial (ICP) – a denominator of a closed-loop transfer function with interval parameters. Due to interval uncertainty of ICP coefficients its roots, which are also poles of a system, migrate inside some regions of a complex plane. Consequently, root quality indices – stability degree and oscillability degree – vary too. That leads us to using robust root quality indices: robust stability degree and robust oscillability degree. Values of these indices can be found from borders of ICP roots allocation area [5]-[9].

In case of interval uncertainty of ICP coefficients values of robust quality indices can be found from allocation of images of coefficients polytope vertices. Quantity of vertices of ICP coefficients polytope depends on a quantity of interval coefficients of ICP. Consequently, reducing a number of vertices, which determine a control quality, is a relevant problem for ICS with numerous interval parameters [10] – [11].

2. Interval extension of expressions for departure angle from complex-conjugate poles

Let us write a general form of an ICP as follows:

$$A(s)=\sum_{i=0}^{n}[a_i]s^i = [a_n]s^n+[a_{n-1}]s^{n-1}+\ldots+[a_0],\ [a_i]>0, \quad (1)$$

where $m = n+1$ coefficients may vary within known intervals. All these coefficients form a polytope M. Let us designate vertices of M as follows: V_q, $q = \overline{1, 2^m}$. Let us notice, that every vertex of M corresponds to a characteristic polynomial with constant parameters, which has, consequently, constant roots.

Coordinates of any point inside M relatively to vertex V_q, $q = \overline{1, 2^m}$ can be calculated via following expressions:

$$a_i = a_i^q + \Delta a_i, i = \overline{0, n} \tag{2}$$
$$(a_{i\min} - a_i^q) \le \Delta a_i \le (a_{i\max} - a_i^q),$$

where Δa_i –is an increment of i-th interval coefficient, a_i^q – is a value of i-th interval coefficient in a vertex V_q. Upper limit of a coefficient a_i is usually designated as $\overline{a_i}$; a lower limit of the coefficient a_i is designated as $\underline{a_i}$. An expression, which links points of M with ICP roots can be obtained via substituting (2) to (1):

$$A^q(s) + \sum_i \Delta a_i s^i = 0,\ i = \overline{0, n}, \tag{3}$$

where $A^q(s) = \sum_i a_i^q s^i$ – is a vertex polynomial. Let us consider edges of M, designated as R_i^q, where i – is an index of interval coefficient, q – is an index of a vertex, from which an edge of departs a_i. On a base of (3) let us write an equation of R_i^q image on a complex plane:

$$A^q(s) + \Delta a_i s^i = 0. \tag{4}$$

According to root locus theory equation (4) can be obtained from an edge transfer function:

$$W_i^q(\Delta a_i, s) = \frac{\Delta a_i s^i}{A^q(s)}. \tag{5}$$

From analysis of (4) and (5) it can be concluded, that variation of Δa_i in an interval (3) leads to roots migration from poles of (5), which correspond to one end of R_i^q, to roots (4) on the other end. This migration corresponds to edge branches of a root locus RS_i^q, its beginnings and ends – to root nodes U_q.

According phase equation of a root locus [12], a angle of a RS_i^q departing from a complex pole P_1, which is a root node U_q, with increasing a_i

$$\Theta_1^i = \pi - \sum_{p=2}^{n} \Theta_p + i\Theta_0, \tag{6}$$

with a_i decreasing

$$\Theta_1^i = -\sum_{p=2}^{n} \Theta_p + i\Theta_0, \tag{7}$$

where Θ_p and Θ_0 –are angles between a real axis and vectors, directed from P_1 to p-th pole and to i-th zeros with coordinates $(0; j0)$.

Let us extend (6) and (7) to interval numbers. Assume, that robust oscillability degree of a considered system is determined by a pair of complex-conjugate poles P_1 and P_2 (figure 1).

IOP Conf. Series: Materials Science and Engineering **707** (2019) 012013 doi:10.1088/1757-899X/707/1/012013

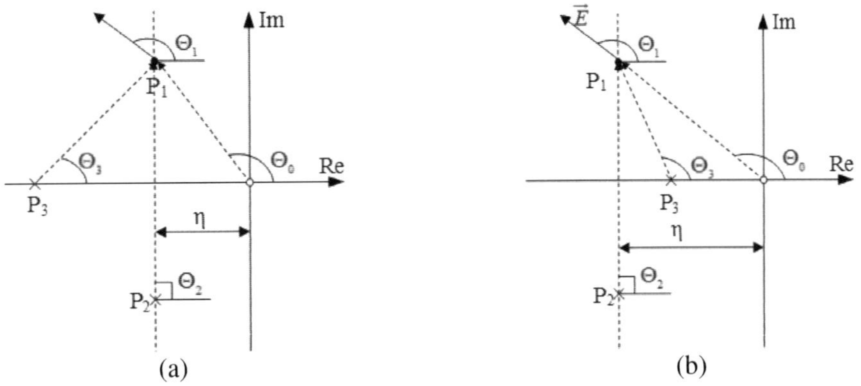

Figure 1. Real pole P_3 allocation a) on the left from P_1 and P_2; b) on the right of P_1 and P_2.

Considering an ability of P_1 to change its location, an angle Θ_0 will have interval value, designated as $[\Theta_0]$. Allocation area of P_1 lies in the second quadrant, so $[\Theta_0]=[90°;180°]$. Other angles Θ_p will also have interval values $[\Theta_p]$.

Let us now derive an expression for $[\Theta_p]$ in two practically significant cases: when a pair of complex-conjugate poles P_1 and P_2 lies on the left or on the right side of a real pole P_3 (figure 1a and 1b). Let us notice, that $\Theta_2 = 90^{\circ}$.

Statement 1. If a real pole P_3 migrates inside an area, which is located on the left of a complex-conjugate poles P_1 and P_2 allocation area (figure 1a), then angle $[\Theta_3]$ of pole P_3 relatively to pole P_1 varies within the range $[0^{\circ};90^{\circ}]$.

Proof. If P_3 is located on the left of the line, going through P_1 and P_2, and near it, then $\Theta_3 \to 90^{\circ}$. If P_3 is located far enough away of the line, then $\Theta_3 \to 0^{\circ}$. Consequently, $[\Theta_3]=[0^{\circ};90^{\circ}]$.

Statement 2. If the real pole P_3 migrates in inside an area located on the right of complex-conjugate poles P_1 and P_2 allocation area (figure 1b), then angle of P_3 relatively to P_1 varies within the range $[90^{\circ};\Theta_0]$.

Proof. If P_3 is located on the right of a line, going through P_1 and P_2, and on the right of it, then $\Theta_3 \to 90^{\circ}$. Moving P_3 away of the line to the right is limited by the origin of coordinates, so $\Theta_3 \to \Theta_0$. Consequently, $[\Theta_3]=[90^{\circ};\Theta]$.

Considering these two statements an interval extension of (6) and (7) resulted in following expressions:

$$[\Theta_1^i] = \pi - \sum_{p=2}^{n}[\Theta_p] + i[\Theta_0] \quad (8)$$

$$[\Theta_1^i] = -\sum_{p=2}^{n}[\Theta_p] + i[\Theta_0] \quad (9)$$

3. Interval phase inequalities, determining a robust oscillability degree of a system

If a pole P_1 determines a robust stability degree, then all other poles lie in a sector $\varphi = \pm\Theta_0$. So, it is possible to constrain departure angles $[\Theta_1^i]$ of edge branches from a pole P_1

$$[\Theta_0] \le [\Theta_1^i] \le [\Theta_0] + \pi, i = \overline{0,n}. \quad (10)$$

Let us notice, that in order to satisfy (10) for all of ICP coefficients a summand πr_i, where $r_i = 0$ or $r_i = 1$ should be added to (8) and (9). Choosing a value of $r_i = 1$ adds 180° to a departure angle along the edge of a_i; choosing a $r_i = 0$ do not change a departure angle. Adding such summand enables us to consider both lower and upper limits of each coefficient a_i. Considering this, let us rewrite (10) as follows:

$$[\Theta_0] \le [\Theta_0]\cdot i - \sum_{p=2}^{n}[\Theta_p] \pm \pi r_i \le [\Theta_0] + \pi, i = \overline{0,n}. \quad (11)$$

Let us designate inequalities (11) as interval phase inequalities.

4. Finding verifying vertices for estimating a robust oscillability degree

On a basis of statements 1 and 2, let us find verifying vertices of a coefficients polytope for ICS of the second and the third order.

Statement 3. Robust oscillability degree of an ICS of the second order can be found through calculating roots of its ICP in a following vertex of a coefficients polytope: $V_1 = \overline{a_0}\,\underline{a_1}\,\overline{a_2}$.

Proof. Let us write interval phase inequalities for ICS of the second order:

$$[\Theta_0] \le [\Theta_0]\cdot(i-1) \pm \pi r_i \le [\Theta_0] + \pi, i = \overline{0,2}. \quad (12)$$

By solving it, we obtain that *r*=0 for *i*=0; *r*=1 for *i*=1 and *r*=0 for *i*=2. Consequently, coordinates of the verifying vertex can be written as follows: $V_1 = \overline{a_0}\,\underline{a_1}\,\overline{a_2}$.

Statement 4. Robust oscillability degree of an ICS of the third order can be found by examining following vertices of an ICP coefficients polytope: $V_1 = \overline{a_0}\,\underline{a_1}\,\overline{a_2}\,\overline{a_3}$, $V_2 = \overline{a_0}\,\underline{a_1}\,\underline{a_2}\,\overline{a_3}$, $V_3 = \overline{a_0}\,\underline{a_1}\,\overline{a_2}\,\underline{a_3}$, $V_3 = \overline{a_0}\,\underline{a_1}\,\underline{a_2}\,\overline{a_3}$, $V_4 = \overline{a_0}\,\overline{a_1}\,\underline{a_2}\,\overline{a_3}$.

Proof. Let us consider the case, when a real pole P_3 lies on the left from the pair of complex-conjugate poles P_1 and P_2. In this case, on a basis of statement 1, interval phase inequalities can be written as follows:

$$90^0 \le [90^0;180^0]\cdot(i-1) - [0^0;90^0] \pm \pi r_i \le 270^0, i = \overline{0,3}.$$

From this expression, a following expression can be obtained:

$$[\Theta_0] \le [90^0(i-1) - 90^0;180^0(i-1)] \pm \pi r_i \le [\Theta_0] + \pi, i = \overline{0,3} \quad (13)$$

Solution of (13) are following values: $r_0 = 0$; $r_1 = 1$; $r_2 = 0$ or $r_2 = 1$, $r_3 = 0$ or $r_3 = 1$. These values correspond to following vertices of a coefficients polytope: $V_1 = \overline{a_0}\,\underline{a_1}\,\overline{a_2}\,\underline{a_3}$, $V_3 = \overline{a_0}\,\underline{a_1}\,\overline{a_2}\,\overline{a_3}$, $V_3 = \overline{a_0}\,\underline{a_1}\,\underline{a_2}\,\overline{a_3}$. In the second case, if a real pole P_3 lies on the right from a pair of complex-conjugate poles P_1 and P_2, then interval phase inequalities can be written as follows on a basis of statement 2:

$$[\Theta_0] \le [90^0;180^0]\cdot(i-1) - [90^0;180^0] \pm \pi r_i \le [\Theta_0] + \pi, i = \overline{0,3}. \quad (14)$$

As a result of (14), interval phase inequalities can be written as follows:

$$90^0 \le [90^0(i-1) - 180^0;180^0(i-1) - 90^0] \pm \pi r_i \le 270^0, i = \overline{0,3} \quad (15)$$

$$90^0 \le [90^0;180^0]\cdot(i-2) \pm \pi r_i \le 270^0, i = \overline{0,3} \quad (16)$$

According to [13] – [14], if there are several ways to write an expression with interval numbers, a correct way of writing an expression is the one, which provides the narrowest interval value. Consequently, a solution of (15) and (16) consists in following vertex coordinates: $V_4 = \overline{a_0}\,\overline{a_1}\,\underline{a_2}\,\overline{a_3}$.

5. Example

5.1. Example 1

Let us consider characteristic polynomial of second order ICS: $[a_2]\cdot s^2 + [a_1]\cdot s + [a_0] = 0$, where $[a_0] = [8,5;11,5]$, $[a_1] = [14;18]$, $[a_2] = [6;8]$.

It is necessary to find a robust oscillability degree of the ICS. To do this, according to developed analysis method let us find roots of the ICP in a following vertex of a coefficients polytope: $V_1 = \overline{a_0}\,\underline{a_1}\,\overline{a_2}$. In this vertex of a coefficients polytope considered ICP has following roots: $s_1 = -0,875 - 0,819j$, $s_2 = -0,875 + 0,819j$; consequently, a robust oscillability degree is equal to $\mu = 0,93$, $\varphi = \pm 43^{\circ}$.

In order to prove the values of a robust oscillability degree, obtained via proposed analysis method, an interval root locus was plotted for considered system (figure 2). Robust oscillability degree of the system is equal to 0.93 and it is the same value, which was found via proposed method. From the figure 2 it is clear, that roots of the considered ICP, which determine a robust oscillability degree of the system, correspond to the vertex $V_1 = \overline{a_0}\,\underline{a_1}\,\overline{a_2}$, which was found previously via proposed method.

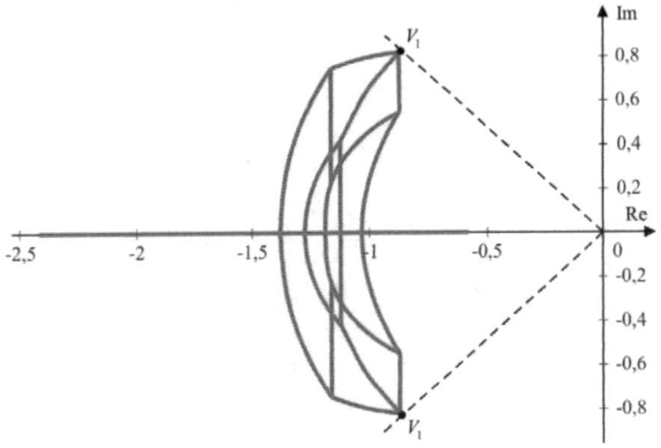

Figure 2. Roots allocation areas

5.2. Example 2

Let us consider characteristic polynomial of second order ICS $[a_3]\cdot s^3 + [a_2]\cdot s^2 + [a_1]\cdot s + [a_0] = 0$, where $[a_0] = [8;10]$, $[a_1] = [6,64;6,94]$, $[a_2] = [1,72;1,95]$, $[a_3] = [0,15;0,5]$. It is necessary to find a robust oscillability degree of the ICS. For its analysis, the root values of the ICP at the verifying vertex were found and are given in Table 1.

Table 1. Values of vertex polynomials roots and robust oscillability degree

Vertex	Values of roots			
	s_1	s_2	s_3	μ_{max}, φ
$V_1(s)$	$-0,9-2,9j$	$-0,9+2,9j$	-2.1	3.303, 73°
$V_2(s)$	$-0,75-3,1j$	$-0,75+3,1j$	-1.9	0.865, 41°
$V_3(s)$	$-2 - 1.8j$	$-2 + 1.8j$	-8.8	4.173, 76°
$V_4(s)$	$-0,8-3,2j$	$-0,8+3,2j$	$-1,8$	3.76, 75°

From the table 1 it is clear, that robust oscillability degree value is determined by roots $s_{1,2} = -0,755 \pm 3,13j$, which correspond to the vertex $V_2 = \overline{a_0}\,\underline{a_1}\,\underline{a_2}\,\overline{a_3}$. Interval root locus of the system is shown in the figure 3. It shows, that robust oscillability degree is equal to $\mu = 4.173$, $\varphi = \pm 76^{\circ}$ and is determined by roots $s_{1,2} = -0,755 \pm 3,13j$ and the vertex $V_2 = \overline{a_0}\,\underline{a_1}\,\underline{a_2}\,\overline{a_3}$.

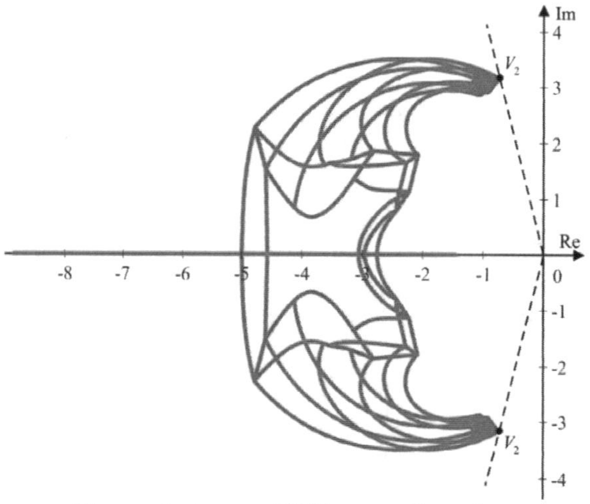

Figure 3. Areas of ICP roots allocation

6. Conclusion

The research resulted in an interval extension of angular equation from a root locus theory. On its basis interval angular inequalities, allowing to find a robust oscillability degree of an ICS through estimating departure angles of root locus edge branches from vertices of ICP coefficients polytope. Solving these inequalities allowed finding sets of vertices, which must be examined to find an oscillability degree of an ICS. Examples of analyzing a control quality of a control system with interval parametric uncertainty prove correctness of the proposed analysis method.

7. Acknowledgments

The reported study is supported by the Ministry of Education and Science of Russian Federation (project #2.3649.2017/PCh).

8. References

[1] Husek P 2008 Robust Stability of Polynomic Interval Polynomials *Proc. of the 17th World Cong. The Int. Federation of Automatic Control (Seoul)* 5862-5867.

[2] Kostov K, Karlova V and Todorov A 2008 Robust root locus application in design and analysis of typical industrial control system model, *Cybernetics and Information Technologies* **8(1)** 25-33.

[3] Tong Y and Sinha N 1994 A computational technique for the robust root locus *IEEE Trans. on Industrial Electronics* **41(1)** 79-85.

[4] Rimsky G and Nesenchuk A 1996 Root locus methods for robust control systems quality and stability investigations *Proc. of IFAC 13th Triennial World Congress (San Francisco)* 469-474.

[5] Nesenchuk A 2010 The root-locus method of synthesis of stable polynomials by adjustment of all coefficients *Automation and Remote Control* **71(8)** 1515-1525.

[6] B.T. Polyak, P.S. Scherbakov, *Robust stability and control*, p.303 (Science, Moscow, 2002).

[7] Wang Y and Hunt K 2000 The calculation of stability radius with D-stability region and non-linear coefficients *Proc. of 3rd IFAC Symp. on Robust Control Design (Prague)* 240-246.

[8] Wang Z, Wang L and Yu W Determinative vertices of interval family with Ω *J. of Mathematical Analysis and Applications* **266(2)** 321-332.

[9] Vukosavich S and Stoich M 1996 Sufficient conditions for the robust relative stability of linear continuous systems *Automation and Remote Control* **57(11)** 84-91.

[10] Gayvoronskiy S, Ezangina T, Khozhaev I and Gunbo L 2017 The analysis of the root quality factors of a power unloading system *Journal of Physics: Conf. Series* **803(1)** 012044.

[11] Vadutov O and Gaivoronskii S 2003 Application of edge routing to stability analysis of interval polynomials *J. of Computer and Systems Sciences Int.* **42(6)** 833-838.

[12] Uderman E 1972 *Root Locus Method in Control Theory* (Moscow: Science).

[13] Shary S 1995 Solving the linear interval tolerance problem *Mathematics and Computers in Simulation* **39 (1-2)** 53-85.

IOP Conf. Series: Materials Science and Engineering **707** (2019) 012013 doi:10.1088/1757-899X/707/1/012013

[14] Shary S 2015 New characterizations for the solution set to interval linear systems of equations *Applied Mathematics and Computation* **265** 570-73.

IOP Conf. Series: Materials Science and Engineering **707** (2019) 012014 doi:10.1088/1757-899X/707/1/012014

Comparative study of three robust observers for automotive damper force estimation

Thanh-Phong Pham[1, 2], Olivier Sename[1] and Luc Dugard[1]

[1] Univ. Grenoble Alpes, CNRS, Grenoble INP, GIPSA-lab, 38000 Grenoble, France. Institute of Engineering Univ. Grenoble Alpes.
[2] Faculty of Electrical and Electronic Engineering, The University of Danang - University of Technology and Education, 550000 Danang, Vietnam

Email: {thanh-phong.pham2; olivier.sename; luc.dugard}@gipsa-lab.grenoble-inp.fr

Abstract. This paper aims at comparing three robust observers used to estimate the damping force of electrorheological (ER) dampers in a vehicle suspension system. Firstly, a nonlinear quarter-car model, augmented with a first-order dynamical nonlinear damper model, is developed. The first two methods are designed considering the nonlinearity as an unknown input and minimizing the effect of the unknown input disturbances (including a nonlinearity term, the measurement noise and the unknown road profile) on the estimation errors, by using an H_2 and H_∞ criterion, respectively. The latter method aims at minimizing only the effects of measurement noises and road profiles on the state variable estimation errors by using a H_∞ criterion, while the nonlinearity is bounded through a Lipschitz condition. For implementation issue, two low-cost sensors signals (two accelerometers data from the sprung mass and the unsprung mass) are considered as inputs for the observer designs. Then, the observers are implemented in real-time on the INOVE test bench from GIPSA-lab (1/5-scaled real vehicle) to assess and compare experimentally the performances of the approaches. Both simulations and experimental results demonstrate a better effectiveness of the latter observer in terms of the ability of estimating the damper force in real-time despite the nonlinearity, the measurement noises and the road disturbances.

Keywords: *Semi-active suspension, nonlinear model, H_2 observer, H_∞ observer, damper force estimation.*

1. Introduction

It is now well known that semi-active suspensions are a key component of vehicles, with the aim of improving comfort and safety (road holding) for on-board passengers, see [1]. The design of control algorithms for automotive semi active suspension system has received a lot of consideration in the last 20 years. Many approaches are considering the damper force as the control input of the vehicle model, and then use, for implementation, an inverse damper model (or look-up tables) to compute the input current or voltage (see [2, 3]). Few papers use local force tracking control schemes in order to attain control objectives (see [4]). However, as damper force sensors are difficult to install and expensive setups, the real-time estimation of the damper force is crucial for suspension control (and reliability). To this aim, some specific methods have been developed to estimate the damper force (see [5–10]). The key challenges are to reduce the cost of the required sensors, to take the non linear dynamical behavior of the damper into account, in presence of unknown road disturbance and sensor noises.

This paper presents a comparative study of three robust control approaches for solving the above challenges in designing the damper force observer of the semi-active suspension systems. The aspects of comparison are a) dealing with nonlinearity, b) minimizing the effect of sensor noise and unknown road profile to estimation errors. Note that such a comparison is proposed here for the first time in the case of suspension system. All of the observers use only two accelerometers in order to estimate the damper force in the quarter-car vehicle equipped with ER suspension. The designs of the observers are based on a nonlinear suspension model consisting of a quarter-car vehicle model, augmented with a first order dynamical nonlinear damper model. To handle the nonlinearity in the damper model, the two first observers (H_2 and H_∞ observers) consider the nonlinearity as a unknown disturbance and use the H_2 and H_∞ norms to minimize the effect of three unknown input disturbances (nonlinearity, sensor noises, unknown road profile) on the estimation errors of the state variables, respectively. In the latter observer (H_∞ Lipschitz observer), the nonlinearity is bounded by a Lipschitz condition, while the H_∞ norm is utilized to minimize the effect of sensor noises and road profile disturbance on the estimation errors. To assess and compare experimentally the performances of the observers, they have been implemented on a real scaled-vehicle test bench, through the Matlab/Simulink real-time workshop.

The remainder of this paper is organized as follows. Section II presents semi-active suspension modeling and III the design of the observers. In section IV, the analysis and comparison of the observer designs are provided in the frequency and time domains. Section V shows the experimental validation and comparison. Section VI gives some concluding remarks.

2. Semi-active suspension modeling and quarter-car system description

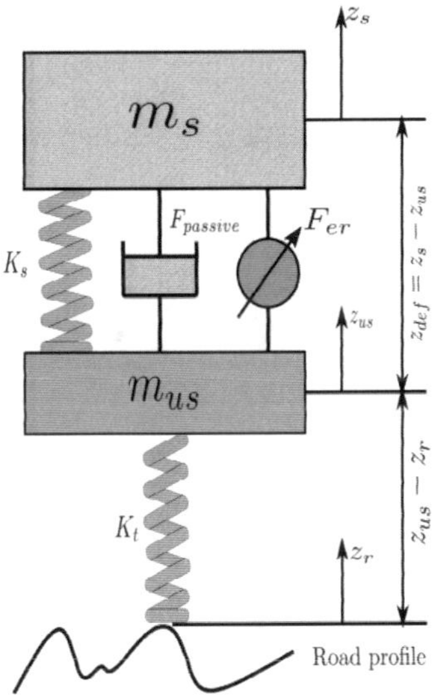

Figure 1. 1/4 car model with semi-active suspension

This section introduces the nonlinear quarter-car model equipped with the semi-active ER suspension system. The quarter-car model equipped with the semi-active ER suspension system, depicted in Fig.1, is described in this section. The well-known model consists of the sprung mass (m_s), the unsprung mass (m_{us}), the suspension components located between m_s and m_{us} and the tire which is modelled as a spring with stiffness k_t. From second law of Newton for motion, the vertical system dynamics around the equilibrium are given as:

$$\begin{cases} m_s\ddot{z}_s = -F_s - F_d \\ m_{us}\ddot{z}_{us} = F_s + F_d - F_t \end{cases} \tag{1}$$

where $F_s = k_s(z_s - z_{us})$ is the spring force; $F_t = k_t(z_{us} - z_r)$ is the tire force; the damper force F_d is given as follows [10]:

$$\begin{cases} F_d = k_0(z_s - z_{us}) + c_0(\dot{z}_s - \dot{z}_{us}) + F_{er} \\ \dot{F}_{er} = -\frac{1}{\tau}F_{er} + \frac{f_c}{\tau}.u.\tanh(k_1(z_s - z_{us}) + c_1(\dot{z}_s - \dot{z}_{us})) \end{cases} \quad (2)$$

where k_0, k_1, c_0, c_1, f_c, τ are constant parameters; z_s and z_{us} are the displacements of the sprung and unsprung masses, respectively; z_r is road displacement input.

By select the system states as $x = [x_1, x_2, x_3, x_4, x_5]^T = [z_s - z_{us}, \dot{z}_s, z_{us} - z_r, \dot{z}_{us}, F_{ER}]^T \in \mathbb{R}^5$, the variables to be estimated $z = [x_1, x_2, x_3, x_5]^T \in \mathbb{R}^4$, the measured variables $y = [\ddot{z}_s, \ddot{z}_{us}]^T \in \mathbb{R}^2$ and the control input $u \in \mathbb{R}$, the system dynamics in the state-space representation can be written as follows:

$$\begin{cases} \dot{x} = Ax + B\Phi(x).u + D_1\omega \\ y = Cx + D_2\omega \\ z = C_z x \end{cases} \quad (3)$$

where $\Phi(x) = \tanh(k_1x_1 + c_1(x_2 - x_4))$; $\omega = \begin{pmatrix} \dot{z}_r \\ n \end{pmatrix}$, in which $\dot{z}_r$ is road profile derivative, n is measurement noises; the system matrices A, B, C, D_1, D_2, C_z are known.

Note that the measured outputs $y = [\ddot{z}_s, \ddot{z}_{us}]^T$ can be obtained easily from on board sensors (accelerometers).

3. Observer design

In this section, three observers (H_2, H_∞ and H_∞ Lipschitz observers) are developed to estimate the damping force accurately. In the design step, the sensor noises and road profile are considered as the unknown input ω. Therefore, H_2 and H_∞ observers are designed to minimize the effect of the unknown disturbance ω and of the nonlinearity on the state estimation errors, while an H_∞ Lipschitz observer is developed to minimize the effect of the unknown input ω on the state estimation errors bouding the nonlinearity by Lipschitz constant.

Firstly, H_2, H_∞ and H_∞ Lipschitz observers for the system (3) are defined in same form as follows:

$$\begin{cases} \dot{\hat{x}} = A\hat{x} + L(y - C\hat{x}) + B\Phi(\hat{x}).u \\ \hat{z} = C_z\hat{x} \end{cases} \quad (4)$$

where $\hat{x}$ is the estimated states vector of x. $\hat{z}$ represents the estimated variables of the variables z. The observer gain L will be determined in the next subsections. The estimation error is given as:

$$e(t) = x(t) - \hat{x}(t) \quad (5)$$

Differentiating $e(t)$ with respect to time and using (3) and (4), leads to:

$$\begin{cases} \dot{e} = \dot{x} - \dot{\hat{x}} = (A - LC)e + B\big(\Phi(x) - \Phi(\hat{x})\big).u + (D_1 - LD_2)\omega \\ e_z = C_z e \end{cases} \quad (6)$$

Remark 1: In order to distinguish among three methodologies, the observer gain L is denoted as L_2 for the H_2 observer, L_∞ for the H_∞ observer and $L_{\infty L}$ for the H_∞ Lipschitz observer.

3.1. H_2 observer design

As discussed previously, an H_2 observer is designed to minimize the effect of ω and $\big(\Phi(x) - \Phi(\hat{x})\big).u$ on the state estimation errors e_z.

From (6) and Remark 1, one obtains

$$\begin{cases} \dot{e} = (A - L_2C)e + D\omega_n \\ e_z = C_z e \end{cases} \quad (7)$$

where $D = (B \;\; (D_1 - L_2D_2))$, $\omega_n = \begin{pmatrix} \big(\Phi(x) - \Phi(\hat{x})\big).u \\ \omega \end{pmatrix}$.

The transfer matrix between the state estimation error e_z and the unknown disturbance ω_n is:

$$T_{e_z\omega_n}(s) = C_z\big(sI - (A - L_2C)\big)^{-1}D \quad (8)$$

The H_2 observer design objectives are the following ones:

- The system (7) is stable for $\omega_n = 0$.

- $\left\|T_{e_z\omega_n}(s)\right\|_2$ is minimized for $\omega_n \neq 0$.

The following theorem solves the above problem into an LMI framework [11].

Theorem 1: Consider the system model (3) and the observer (4). If there exist a symmetric positive definite matrix P_2 and a matrix Y_2 minimizing γ_2 such that:

$$\begin{pmatrix} P_2A + A^TP_2 - Y_2C - (Y_2C)^T & P_2B & P_2D_1 - Y_2D_2 \\ B^TP_2 & -I & 0 \\ D_1^TP_2 + (Y_2D_2)^T & 0 & -I \end{pmatrix} < 0$$

$$\begin{pmatrix} P_2 & C_z^T \\ C_z & \gamma_2 I \end{pmatrix} > 0 \tag{9}$$

then, the observer gain L_2 determined from $L_2 = P_2{}^{-1}Y_2$ ensures that the objectives are attained.

3.2. H_∞ observer design

Similarly to subsection 3.1, an H_∞ observer is designed to minimize the effect of ω and $(\Phi(x) - \Phi(\hat{x})).u$ on the state estimation errors e_z.

From (6) and Remark 1, one obtains

$$\begin{cases} \dot{e} = (A - L_\infty C)e + D\omega_n \\ e_z = C_z e \end{cases} \tag{10}$$

where $D = (B \;\; (D_1 - L_\infty D_2))$, $\omega_n = \begin{pmatrix} (\Phi(x) - \Phi(\hat{x})).u \\ \omega \end{pmatrix}$.

The transfer function between the state estimation error e_z and the unknown disturbance ω_n is:

$$T_{e_z\omega_n}(s) = C_z\big(sI - (A - L_2C)\big)^{-1}D \tag{11}$$

The H_∞ observer design objectives are the following ones:

- The system (10) is stable for $\omega_n = 0$.
- $\left\|T_{e_z\omega_n}(s)\right\|_\infty$ is minimized for $\omega_n \neq 0$.

The following theorem solves the above problem into an LMI framework [11].

Theorem 2: Consider the system model (3) and the observer (4). If there exist a symmetric positive definite matrix P_∞ and a matrix Y_∞ minimizing γ_∞ such that:

$$\begin{bmatrix} M & P_\infty B & P_\infty D_1 - Y_\infty D_2 \\ B^TP_\infty & -\gamma_\infty^2 I & 0 \\ D_1^TP_\infty - (Y_\infty D_2)^T & 0 & -\gamma_\infty^2 I \end{bmatrix} < 0 \tag{12}$$

where $M = P_\infty A + A^TP_\infty - Y_\infty C - (Y_\infty C)^T + C_z^TC_z$. Then, the observer gain L_∞ determined from $L_\infty = P_\infty{}^{-1}Y_\infty$ ensures that the objectives are attained.

3.3. H_∞ Lipschitz observer design

In this section, the damping force is estimated through an H_∞ Lipchitz observer whose objectives are to minimize the effects of unknown road profile derivative and measurement noises on the estimation errors of the state variables and to account for the nonlinearity through a Lipschitz assumption.

Firstly, the control input function $\Phi(x)$ of the system (3) can be rewritten under the following form

$$\Phi(x) = tanh\,(\Gamma x) \tag{13}$$

where $\Gamma = [k_1, c_1, 0, -c_1, 0]$.

Therefore, $\Phi(x)$ satisfies the Lipschitz condition in x

$$\|\Phi(x) - \Phi(\hat{x})\| \le \|\Gamma(x - \hat{x})\|, \forall x, \hat{x}. \tag{14}$$

From (6) and Remark 1, the estimation error dynamics of H_∞ Lipschitz observer is obtained

$$\begin{cases} \dot{e} = (A - L_{\infty L}C)e + B\big(\Phi(x) - \Phi(\hat{x})\big).u + (D_1 - L_{\infty L}D_2)\omega \\ e_z = C_z e \end{cases} \tag{15}$$

Assuming the Lipschitz condition (14) for $\Phi(x)$, the H_∞ Lipschitz observer design objective is stated

- The system (15) is stable for $\omega = 0$.
- $\|e_z(t)\|_{\mathcal{L}_2} < \gamma_{\infty L}\|\omega(t)\|_{\mathcal{L}_2}$ for $\omega \neq 0$.

The following theorem solves the above problem into an LMI framework [10].

Theorem 3: Consider the system model (3) and the observer (4). If there exist a symmetric positive definite matrix $P_{\infty L}$ and a matrix $Y_{\infty L}$ minimizing $\gamma_{\infty L}$ such that:

$$\begin{bmatrix} \Omega & P_{\infty L}B & P_{\infty L}D_1 - Y_{\infty L}D_2 \\ * & -\epsilon_l I_d & 0_{n,d} \\ * & * & -\gamma_{\infty L}^2 I \end{bmatrix} < 0 \qquad (16)$$

where $M = P_{\infty L}A + A^T P_{\infty L} - Y_{\infty L}C - (Y_{\infty L}C)^T + \epsilon_l \Gamma^T \Gamma + C_z^T C_z$. Then, the observer gain $L_{\infty L}$ determined from $L_{\infty L} = P_{\infty L}{}^{-1} Y_{\infty L}$ ensures that the objectives are attained.

4. Analysis of the observer design: frequency and time domain simulations

In this section, the synthesis results of the $\mathbf{H_2}$, $\mathbf{H_\infty}$ and $\mathbf{H_\infty}$ Lipschitz observers are presented and some simulation results are provided.

4.1. Synthesis results and frequency domain analysis

Solving theorem 1, theorem 2 and theorem 3 leads the following solutions: the gains $\gamma_2 = 3.1917$, $\gamma_\infty = 1.4142$, $\gamma_{\infty L} = 1.0032$ and the observer gains L_2, L_∞, $L_{\infty L}$.

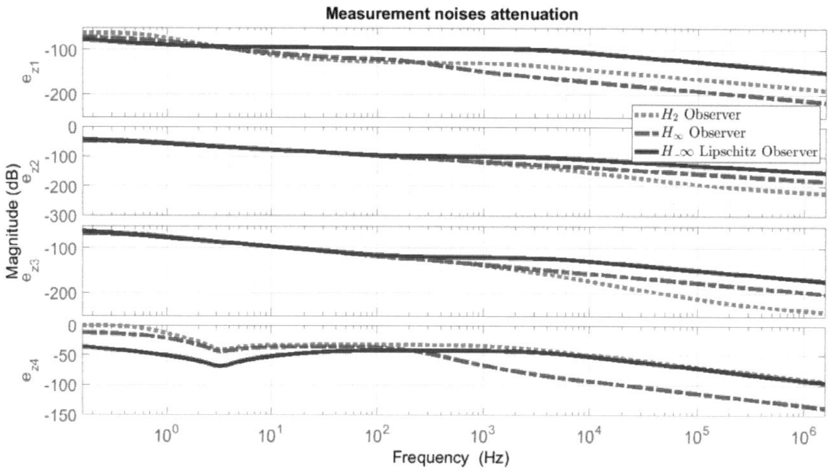

Figure 2. Measurement noises attenuation

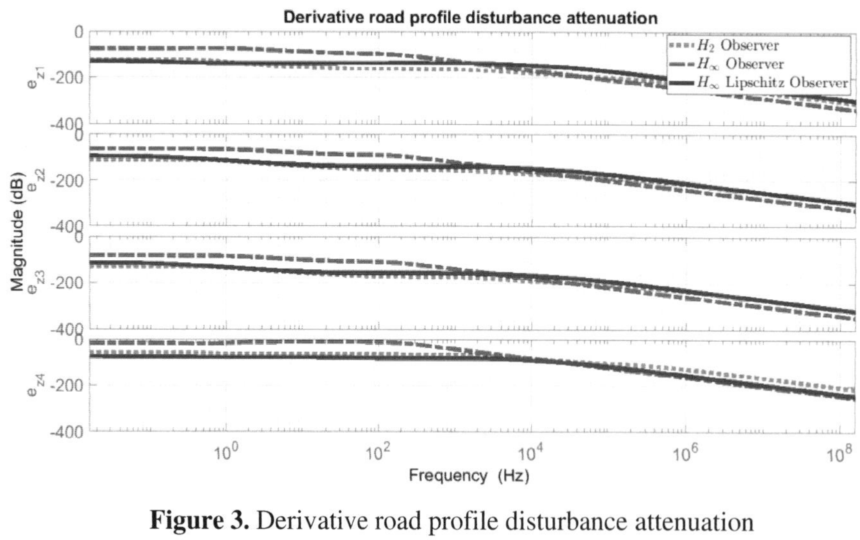

Figure 3. Derivative road profile disturbance attenuation

The resulting attenuation of the sensor noises and derivative road profile disturbance on the estimation error are illustrated in Figures 2 and 3. These figures emphasize the attenuation level of the measurement noises and unknown road profile effect on the 4 estimation errors. The largest sensor noise and derivative road profile disturbance amplification of the 4 errors, over the whole frequency range ([1, 107] Hz for sensor noise and [0.1, 108] Hz for derivative road profile disturbance), are 13dB at 1 Hz and -17.6dB at 0.1 Hz, respectively.

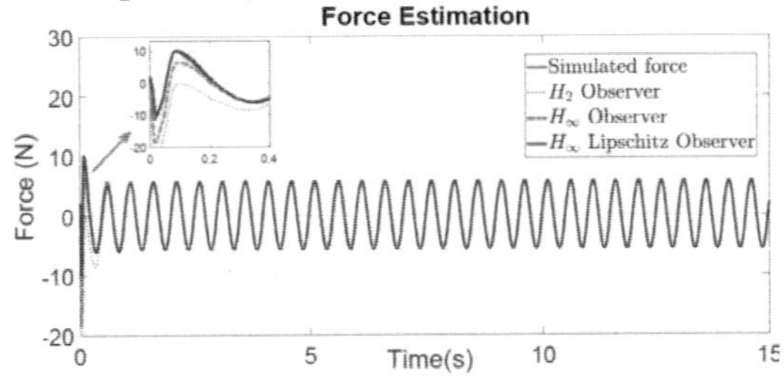

Figure 4. Simulation results: Estimated force

4.2. Simulation

To evaluate the effectiveness of the proposed algorithms, the following simulation is carried out with the initial conditions of the observer states: $\hat{x}_0 = [0.01, 0.1, 0.001, 0.1, 4]^T$ and 0 for the simulated system. The following simulation scenario is considered to compare the performance of the proposed observers: road profile is a sequence of sinusoidal bumps (2Hz) and the control input u is 0.2. The simulation results of three observers are shown in the Fig. 4.

5. Experimental validation

To experimentally assess and compare the effectiveness of the proposed methodologies, the proposed observers are implemented on the 1/5 car scaled car INOVE available at GIPSA-lab, shown in Fig. 5.

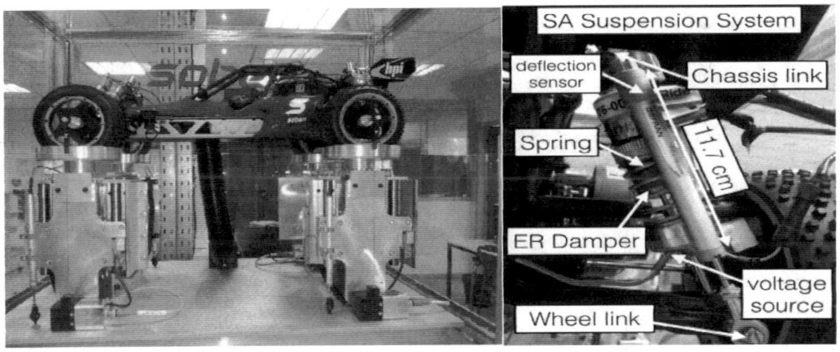

Figure 5. The experimental testbed INOVE at GIPSA-lab (www.gipsa-lab.fr/projet/inove).

This test-bench is equipped with 4 semi-active ER suspensions, which is controlled in real-time using Matlab real-time workshop and a host computer. The three observer system is implemented with the sampling period Ts = 0.005s. Note that the experimental platform is fully equipped with sensors to measure its vertical motion. At each corner of the system, a DC motor is used to generate the road profile.

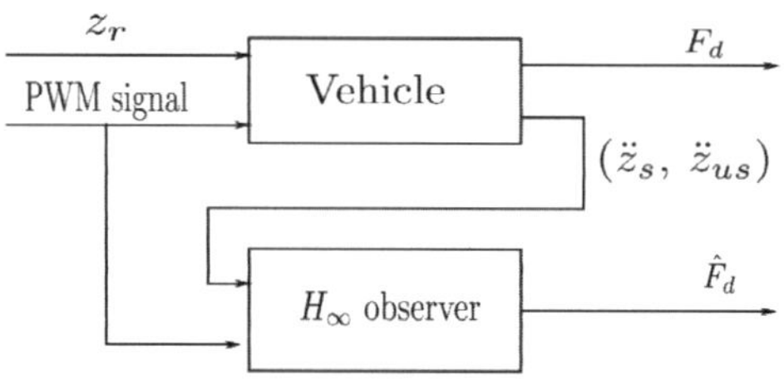

Figure 6. Block diagram for the implementation of the observers.

The observers are applied on the rear-left corner using two accelerometers: the unsprung mass $\ddot{z}_{us}$ and the sprung mass $\ddot{z}_s$. For validation purpose only, the damper force sensor is used to compare the measured force with the estimated one. The following block-scheme illustrates the experimental scenario of the observer (shown in Fig. 6). The experimental scenario is as follows: the road profile is a sequence of sinusoidal bumps and the control input u is constant (u = 0.2).

The experiment results of the observer are presented in Fig. 7. The result illustrates the accuracy and efficiency of the proposed observers. To further describe this accuracy, Table 1 shows the normalized root-mean square errors, considering the difference between the estimated and measured forces in the experiment tests.

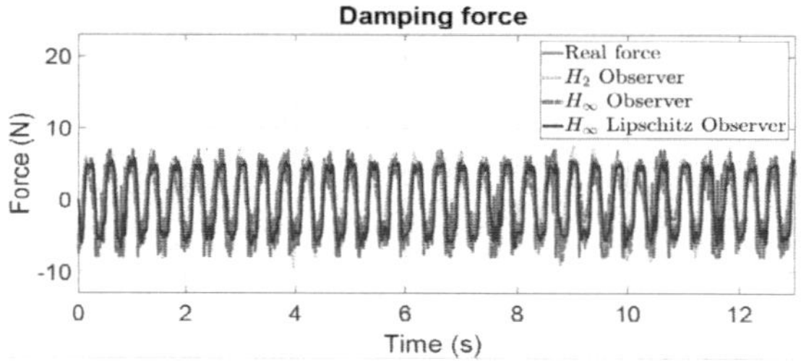

Figure 7. Experimental results of the observers.

Table 1. Normalized Root-Mean-Square Errors (NRMSE).

Observer	NRMSE
H_2 Observer	0.1104
H_∞Observer	0.1032
H_∞ Lipschitz Observer	0.0896

6. Conclusion

This paper developed and compared several observers to estimate the damper force, using a dynamic nonlinear model of the ER damper. For this purpose, the quarter-car system is represented in state-space form by considering a phenomenological model of the damper. Based on two accelerometers, the observers are designed, giving a good estimation result of the damping force. The estimation error is minimized, accounting for the effect of unknown inputs (road profile disturbance and measurement noises) and the nonlinearity term bounded by a Lipchitz condition. Both simulation and experiment results assess the ability and the accuracy of the proposed models to estimate the damping force of the ER semi-active damper.

7. References

[1] S.M. Savaresi, C. Poussot-Vassal, C. Spelta, O. Sename, L. Dugard (2010). *Semi-active suspension control design for vehicles*. Elsevier.

[2] Poussot-Vassal, C., Spelta, C., Sename, O., Savaresi, S. M., & Dugard, L. (2012). Survey and performance evaluation on some automotive semi-active suspension control methods: A comparative study on a single-corner model. *Annual Reviews in Control*, *36*(1), 148-160.

[3] Poussot-Vassal, C., Sename, O., Dugard, L., Gaspar, P., Szabo, Z., & Bokor, J. (2008). A new semi-active suspension control strategy through LPV technique. *Control Engineering Practice*, *16*(12), 1519-1534.

[4] Priyandoko, G., Mailah, M., & Jamaluddin, H. (2009). Vehicle active suspension system using skyhook adaptive neuro active force control. *Mechanical systems and signal processing*, *23*(3), 855-868.

[5] Koch, G., Kloiber, T., & Lohmann, B. (2010, December). Nonlinear and filter based estimation for vehicle suspension control. In *49th IEEE conference on decision and control (CDC)*(pp. 5592-5597). IEEE.

[6] Vela, A. E., Alcántara, D. H., Menendez, R. M., Sename, O., & Dugard, L. (2018). H∞ Observer for Damper Force in a Semi-Active Suspension. *IFAC-PapersOnLine*, *51*(11), 764-769.

[7] Tudon-Martinez, J. C., Hernandez-Alcantara, D., Sename, O., Morales-Menéndez, R., & Lozoya-Santos, J. D. J. (2018). Parameter-Dependent H∞ Filter for LPV Semi-Active Suspension Systems. *IFAC-PapersOnLine*, *51*(26), 19-24.

[8] Rajamani, R., & Hedrick, J. K. (1995). Adaptive observers for active automotive suspensions: theory and experiment. *IEEE Transactions on control systems technology*, *3*(1), 86-93.

[9] Reichhartinger, M., Falkensteiner, R., & Horn, M. (2018). Robust Estimation of Forces for Suspension System Control. *IFAC-PapersOnLine*, *51*(25), 328-333.

[10] Pham, T. P., Sename, O., & Dugard, L. (2019, June). Design and Experimental Validation of an H∞ Observer for Vehicle Damper Force Estimation. In *9th IFAC International Symposium on Advances in Automotive Control (AAC 2019).*

[11] Scherer, C., & Weiland, S. (2000). Linear matrix inequalities in control. *Lecture Notes, Dutch Institute for Systems and Control, Delft, The Netherlands*, *3*(2).

IOP Conf. Series: Materials Science and Engineering **707** (2019) 012015 doi:10.1088/1757-899X/707/1/012015

Peg-in-a-hole by haptic teleoperation

Omer Faruk Argin[1] **and Zeki Yagiz Bayraktaroglu**[2]

[1]Istanbul Technical University, Mechatronics Engineering Department, 34437 MA1 Gumussuyu Beyoglu Istanbul, Turkey
[2]Istanbul Technical University, Mechanical Engineering Department, 34437 MA1 Gumussuyu Beyoglu Istanbul, Turkey

Abstract. This paper presents the experimental study on the application of the peg-in-a-hole task in the proposed haptic teleoperation system. Motion references from the user towards the remote manipulator and force feedback signals from the remote environment towards the user are transmitted through a virtual reality layer where the remote operation is simulated in real-tme. Various physical interactions occuring in the bilatral loop are modelled and controlled through the introduction of virtual spring-damper elements. Modeling and controller design of the teleoperation system is made modular and straightforward with the use of identical virtual dynamics. The design approach combines the introduction of the middle layer of virtual reality and modeling of all interactons by identical virtual spring-dampers. The haptic teleoperation system is experimentally evaluated with experienced and unexperienced users in the bilateral loop. Experimental results show that stable peg-in-a-hole operation is achieved in presence of bounded time-delays in all communication channels between the user, virtual reality layer and remote environment. Improvements of the teleoperation performances in terms of position tracking and total operation time are also observed in experiments.

1. Introduction

Robotic teleoperation is a technique of remote control where a human-operator (user) is involved in the control loop. Actual and future applications of teleoperation systems are found in various fields such as space exploration, nuclear power plants, surgical operations, etc. [1-4]. In bileteral teleoperation of robots, the user transmits motion reference signals to the remote manipulator through convenient physical interfaces. In return, he/she is provided with the information on the state of the remote operation environment by various means of perception such as visual and tactile feedback.
Force feedback is an important information, necessary for satisfactory performance of robotic teleoperation. Therefore, haptic devices with multi-DoFs which provide the user with force feedback from the remote operation are widely used in bilateral teleopration. Development of haptic teleoperation systems in bilateral and multilateral forms is a popular research topic where the stability and transparency are the fundamental performance criteria [5,6].

Time delays on communication channels between local and remote environments of the teleoperation systems have particular impacts on the stability and transparency of the bilateral control loops. Passivity-based and scattering based aproaches [7-9], wave variables [10,11] and intelligent [12,13] methods are proposed in order to improve the stability and transparency against worsening effects of time delays in communication channels.

In our recent work on the design of haptic teleoperation, the proposed control system has been based on the introduction of a virtual haptic coupling between the local and remote sides of the bilateral loop [14]. Virtual interaction forces are then fed back to the user along with visual feedback from the remote operation environment. While the asymptotic stability of the proposed system is

IOP Conf. Series: Materials Science and Engineering **707** (2019) 012015 doi:10.1088/1757-899X/707/1/012015

shown by a Lyapunov analysis, improvements of the performances in terms of position tracking and transparency with the proposed approach is also experimentally demonstrated.

The peg-in-a-hole task, simultaneously combining motion and force controls, is a representative operation for the general problem of assembly [15, 16]. Haptic guidance is a practical approach in trajectory tracking control of remote devices in teleoperation systems [17-20].

In this paper, design and application of a haptic teleoperation system for the peg-in-a-hole task is presented**.** The proposed design is based on an extended use of identical virtual haptic coupling structure which is explained in [14] in the bilateral loop. An additional haptic coupling is proposed in this design. The first virtual coupling is established between the user and remote environment, resulting with virtual interaction forces fed back to the user. Introduction of these virtual interaction forces contribute to the robustness of the teleoperation system against time delays in communication channels.

The second virtual coupling is established between the manipulated peg and the stationary hole, in order to guide the peg for a smooth assembly operation. Guidance of the peg through virtual forces resulting from this second coupling improves the performance of the peg-in-a-hole task by reducing the total operation time.

The expected contributions to the performances of the peg-in-a-hole task by haptic teleoperation are experimentally observed. Stable and transparent operation is achieved in the experiments with experienced and unexperienced users in the bilateral loop. Experimental results show that the proposed haptic coupling design improves the position tracking performance and reduces the interaction forces and total operation time of the peg-in-a-hole assembly.

In the following section, the haptic coupling design proposed for the peg-in-a-hole task is introduced. Section III presents the bilateral control scheme implemented for the haptic teleoperation system. Section IV presents the experimental setup and results with an evaluation of the peg-in-a-hole task. Finally, Section V gives the conclusion of the presented work.

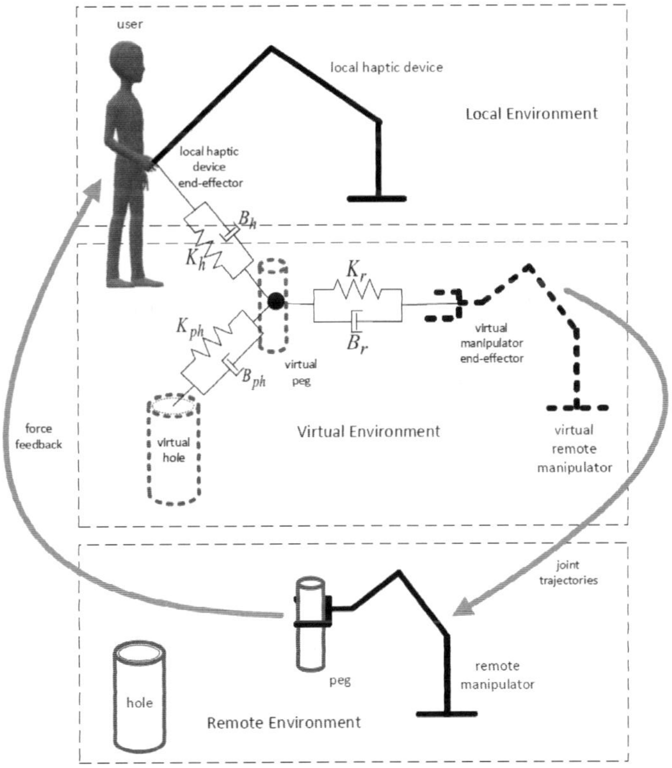

Figure 1. Haptic coupling between the local device and remote manipulator through the virtual reality layer.

2. Haptic Coupling Design

Interacion dynamics occuring in teleoperation systems is usually represented based on the behavior of mechanical spring-damper models. Simplicity of implementation and easy manipulation of these linear second-order nodels lets the designer assign desired impedances in order to control physical interactions. Therefore, it constitutes a viable approach to the problem of haptic coupling between the local and remote sides of bilateral loops [21].

In this work, virtual spring-damper models with identical structure are used for three different purposes as shown in Figure 1. The first purpose is the design of the coupling between the local haptic device and the virtual load, i.e. the peg, visually represented in a real-time simulation environment. Resulting virtual forces from this coupling are fed back to the user in order to induce a feel of inertia, which in turn helps filtering out high frequency motion inputs from the user.

The second coupling is established between the virtual hole and the end-effector of the remote manipulator handling the peg. Resulting virtual forces here are fed to the virtual peg in order to reflect position errors in the remote side of the teleoperation system.

Finally, the third coupling is established between the virtual peg and virtual hole. This coupling is activated when the peg in a hole process approach. Resulting virtual forces are fed back to the user in order to guide the peg-in-a-hole task against varipus disturbances, e.g. modeling errors in the reconstruction of the remote environment in the simulation layer.

The real-time simulation environment constitutes a virtual reality layer between the local and remote sides of the teleoperation system. This layer implemented in an open-source simulation software provides the user with a real-time visual representation of the remote environment during teleoperation. The simulation software computes the forward and inverse kinematics of the remote manipulator as well. Then, in this proposed bilateral scheme, the user directly commands the task space motions necessary to execute the peg-in-a-hole task, without taking care of the manipulator joint trajectories.

The dynamics introduced in the bilateral loop between local haptic device and virtual remote manipulator and the dynamics introduced between the peg and hole is given as follows:

$$m\ddot{x}_{vp} + B_r\dot{x}_{vp} + K_r x_{vp} = B_h\dot{x}_h + K_h x_h \tag{1}$$

$$m\ddot{x}_{vp} + B_r\dot{x}_{vp} + K_r x_{vp} = B_{ph}\dot{x}_{vh} + K_{ph} x_{vh} \tag{2}$$

where x_{vp}, x_{vh} and x_h represent the position vectors of the virtual peg, virtual hole and local haptic device respectively. K_h and B_h are the spring and damper coefficients between the local haptic device and virtual peg, K_r and B_r between the virtual peg and end-effector of the remote manipulator, and K_{ph} and B_{ph} between the virtual peg and virtual hole.

The resulting haptic coupling force vector f_c can be calculated up to the fed coupling type by the Equations 3 and 4.

$$f_c = K_h(x_h - x_{vp}) + B_h(\dot{x}_h - \dot{x}_{vp}) \tag{3}$$

$$f_c = K_{ph}(x_{vh} - x_{vp}) + B_{ph}(\dot{x}_{vh} - \dot{x}_{vp}) \tag{4}$$

The total feedback force vector f_h to be fed back to the haptic interface is then given as follows:

$$f_h = f_c + f_{fbk} \tag{5}$$

where f_{fbk} is the measured remote manipulator's end-effector force vector.

3. Control System Design

The proposed bileteral control scheme describing the teleoperation of the peg-in-a-hole task with a single local haptic device and one remote manipulator is given in Figure 2.

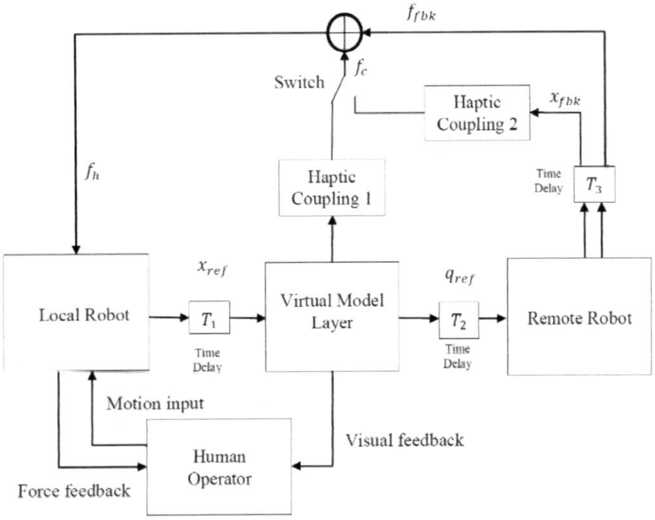

Figure 2. Bilateral control scheme for the peg-in-a-hole task by haptic teleoperation.

A 6-axis industrial robot is used to execute the peg-in-a-hole task. Decentralized PIDs with gravity compensation are implemented as the joint controllers for the motion control of the remote manipulator. The reference joint trajectories are computed by the manipulator model in the virtual reality layer. The PID control law in joint space is given as follows:

$$\tau_i = K_p\left(e_{q_i}(t) + \frac{1}{T_i}\int e_{q_i}(t) + T_d\frac{de_{q_i}(t)}{dt}\right) + \tau_g \tag{6}$$

where K_p , T_i and T_d are the diagonal matrices of PID parameters, τ_g represents the gravity compensation torque, e_q is the vector of joint position errors and i defines joint number. Closed-loop dynamics of the bilateral teleoperation system can be represented as follows:

$$A_m(q_m)\ddot{q}_m + C_m(q_m, \dot{q}_m)\dot{q}_m = \tau_m - \tau_{user} \tag{7}$$

$$A_b(q_b)\ddot{q}_b + C_b(q_b, \dot{q}_b)\dot{q}_b = \tau_b \tag{8}$$

$$A_s(q_s)\ddot{q}_s + C_s(q_s, \dot{q}_s)\dot{q}_s + \tau_g = \tau_s - \tau_e \tag{9}$$

where $A(q)$ is the inertia matrix, $C(q, \dot{q})$ is the matrix of Coriolis and centrifugal forces, τ_g is the vector of torque due to gravity. In above equations 7-9, indices *m*, *b* and *s* denote respectively the local haptic device, virtual model and remote manipulator. $\tau_e$is the vector of interaction force between the remote manipulator and environment, and $\tau_{user}$the generalized force applied by the user through the haptic interface. Control inputs to be applied to the local and remote manipulators are τ_m and τ_s respectively. Here, $\tau_m = J^T f_h$ and τ_s is calculated using the Equation 6.

Asymptotic stability of the proposed bilateral haptic teleoperation system in simple manipulation tasks is shown by a Lyapunov analysis as given in ref. [14].

4. Experiments

4.1. Task Description

The peg-in-a-hole task is achieved in four successive steps as illustrated in Figure 3. In the approach phase, the manipulator freely moves in its workspace in order to pick up the peg from its initial position. The handling phase starts with the grasping of the peg by the manipulator. In this second phase, the manipulator brings the peg in a certain neighborhood of the hole. The haptic coupling 1 is activated in these first two phases of the manipulation. The third phase is the execution of the peg-in-a-hole operation. When the peg becomes sufficiently close to the hole, The bilateral controller switches from the haptic coupling 1 to the haptic coupling 2 (Figure 2. Here, the proposed coupling generates virtual forces along directions normal to the peg and hole axes in order to help aligning them. Finally, the manipulator pulls back the peg from the hole in order to release it in its initial position. Once the peg is separated from the hole, the haptic coupling 1 is reactivated instead of the haptic coupling 2 in the controller.

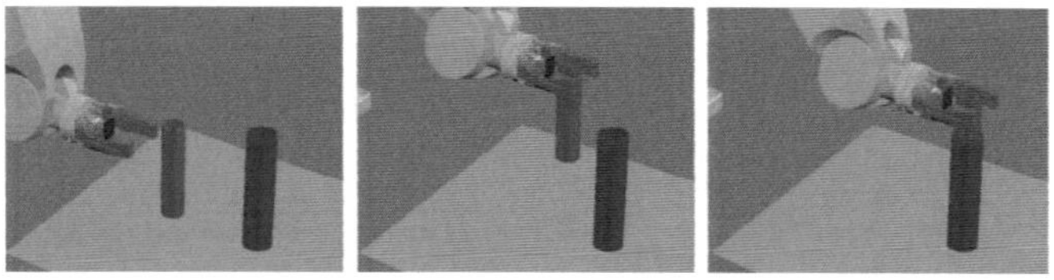

Figure 3. Peg-in-a-hole in the virtual simulation environment a) Approach, b) Handling, c) Peg-in-a-hole.

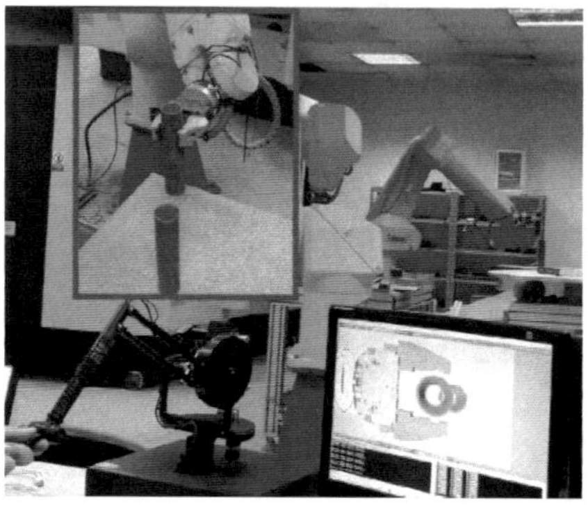

Figure 4. Local haptic device, virtual environment and remote manipulator.

4.2. Experimental Setup

The experimental setup consists of a 6-DoF *Phantom Premium* 1.5 High Force as the local haptic device, a *Stäubli Rx160* [22] robot as the remote manipulator and the *Blender* as the open-source virtual reality software (Figure 4). The remote manipulator is equipped with a 6-axis *ATI* F/T transducer and a single-DoF on/off gripper.

The sampling frequencies of the robot controller, and the haptic interface are 250Hz and 1kHz respectively. The UDP/IP protocol is used in all communications between the local, virtual, and remote environments. Experimentally measured time delays are as follows: $T_1 = 0.7 - 0.8ms$, $T_2 = 0.1 - 0.16ms$, $T_3 = 0.7 - 0.8ms$. The sampling frequency of the robot controller being 250 Hz, these time delays correspond to up to *%20* of one sampling period with T_1 and T_3 and up to *%40* with T_2.

In the proposed bilateral loop, the *Blender* is used for the following purposes: 1) Computation of the forward and inverse kinematics of the remote manipulator, 2) Real-time simulation and visualization of the remote environment for the user, 3) Computation of virtual forces resulting from the proposed haptic couplings.

The use of an open-source software executing the above-mentioned tasks as a middle layer between the local and remote sides of the teleoperation system results in a modular design of the programing environment. Due to the simple structure of the proposed system with *Blender*, kinematic parameters of the remote manipulator and parameters of the virtual haptic couplings can be easily modified by the user. In the case of the remote manipulator being replaced by others, the *Blender* virtual environment offers a user-friendly redesign of the bilateral loop.

4.3. Experimental Results

The proposed teleoperation system is implemented for the peg-in-a-hole task described in in this section. The manipulated peg has the mass $m_p = 0.64kg$, and the position and force scaling factors are implemented as $K_x = 30$ and $K_f = 1/50$ respectively. The parameters of the virtual spring-damper model used in the haptic coupling are $m = 1$, $B_h = 35$, and $K_h = 600$, resulting in the damping ratio $\zeta_h = 0.714$ and natural frequency $\omega_h = 24.5rad/s$. The parameters of the virtual spring-damper designed around the peg-in-a-hole operation are $m = 1$, $B_{ph} = 15$, and $K_{ph} = 1500$, resulting in the damping ratios are $\zeta_{ph} = 0.194$ and natural frequency $\omega_{ph} = 38.73rad/s$.

The PID parameters K_p, T_i, and T_d of the remote manipulator position controller given in Eq. 6 are experimentally tuned as $K_p = [3500\ 8000\ 6500\ 2500\ 1000\ 1000]$, $T_i = [150\ 250\ 250\ 40\ 30\ 30]$ and $T_d = [0.2\ 0.04\ 0.02\ 0.025\ 0.025\ 0.001\ ]$. In order to evaluate the performance of the proposed control system given in Figure 2, two cases are considered in experiments. In the first case, the haptic coupling 1 only is activated during the entire teleoperation. In the second case, the switching between haptic couplings is applied as described in this section.

4.3.1. First Case. In an experimentation with an experienced user in the loop, the tracking performance of the remote manipulator in task space is obtained as given in Figure 5. Figure 6 shows the haptic feedback force f_h computed as defined in Eqs. 3 and 5. Here, the user completes the teleoperation in 66.4s and durations of successive phases are 0-1s: Approach, 1-25s: Handling, 25-52s: Peg-in-a-hole, 52-66 s: Return of the peg to its initial position.

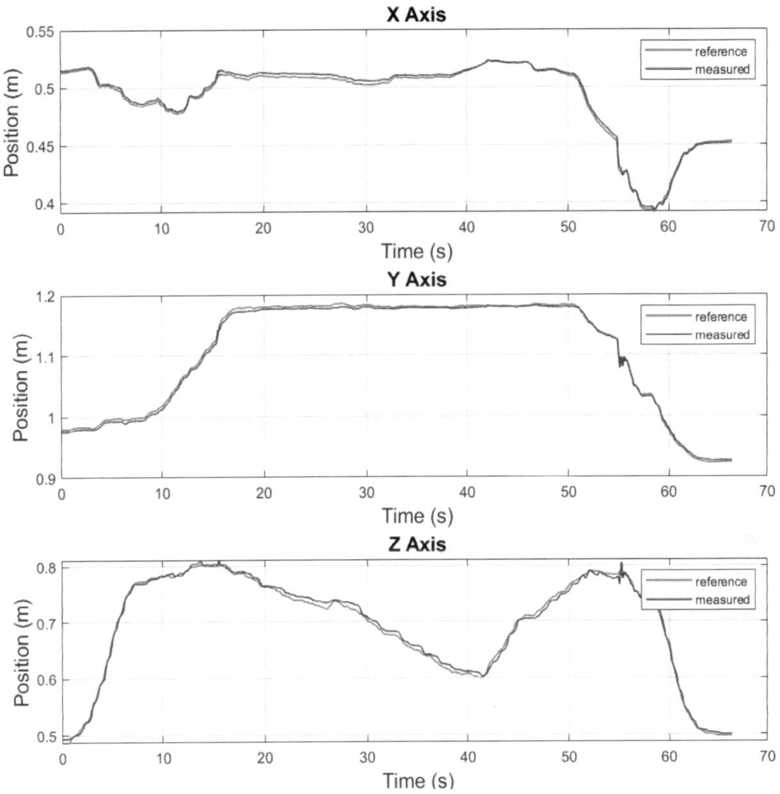

Figure 5. Reference and actual task trajectories of the remote manipulator, with haptic coupling 1.

4.3.2. Second Case. In an experimentation with an experienced user in the loop, the tracking performance of the remote manipulator in task space is obtained as given in Figure 7. Figure 8 shows the haptic feedback force f_h computed as defined in Eqs. 3, 4 and 5. The total operation time is 61.5s and durations of successive phases are 0-1 s: Approach, 1-20 s: Handling, 20-39 s: Peg-in-a-hole, 39 - 61 s: Return of the peg to its initial position.

Experimentations of the peg-in-a-hole task with unexperienced users in the loop are also achieved. Experimental results in terms of considered performance criteria with three unexperienced and one experienced user are given in the following.

Figure 9 shows the position tracking error performances with one experienced and 3 unexperienced users. The average position tracking error and standard deviation of tracking errors are decreased with the introduction of the second haptic coupling. The average position tracking error with unexperienced users is 7.8 mm in the first case, and it drops to 7.5 mm with the introduction of haptic guidance in the second case. Average errors recorded with the experienced user are 7.6 mm and 6.3 mm respectively.

Figure 10 gives the feedback forces applied to the local haptic device. The average feedback force and standard deviation of feedback forces are decreased as well with respect to those observed in the first case. The attenuation of feedback forces means that the peg-in-a-hole task is achieved with smoother physical interaction between the peg and hole. The average feedback force with unexperienced users is 7.4 N in the first case, and it drops to 4.3 N with the introduction of haptic guidance in the second case. Average feedback forces recorded with the experienced user are 6.2 N and 3 N respectively.

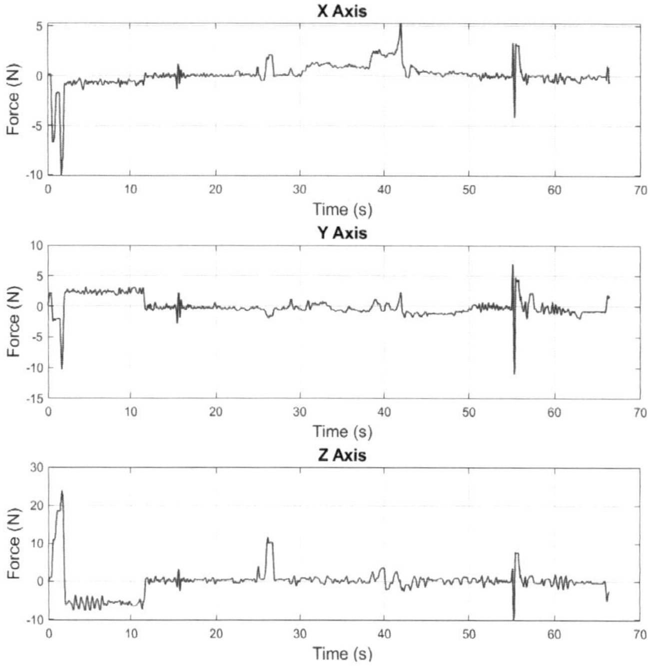

Figure 6. Feedback force (f_h) to the local haptic device, with haptic coupling 1 only.

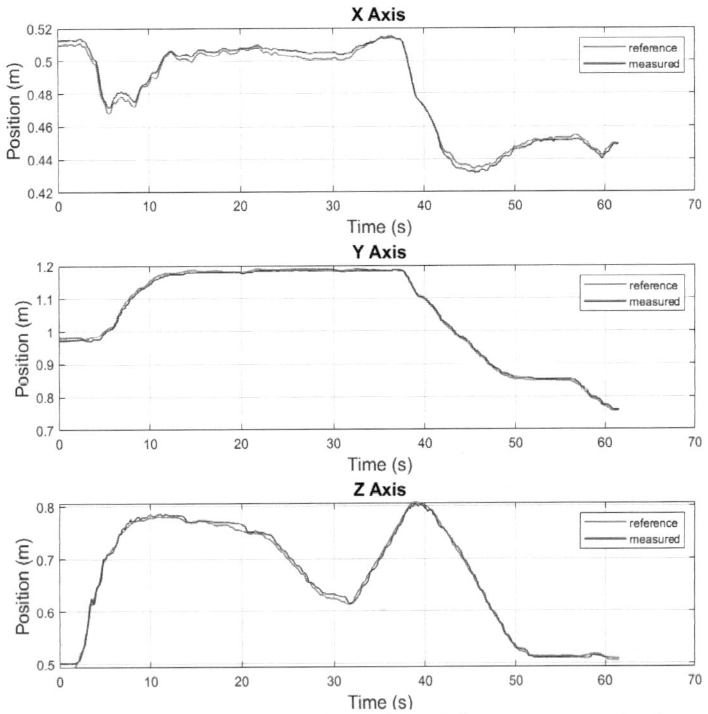

Figure 7. Reference and actual task space trajectories of the remote manipulator, with switchings between haptic couplings.

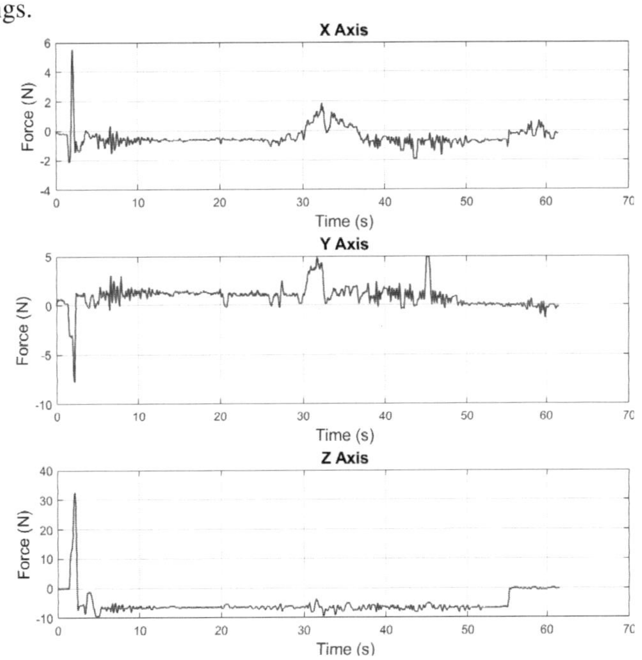

Figure 8. Feedback force (f_h) to the local haptic device, with switchings between haptic couplings.

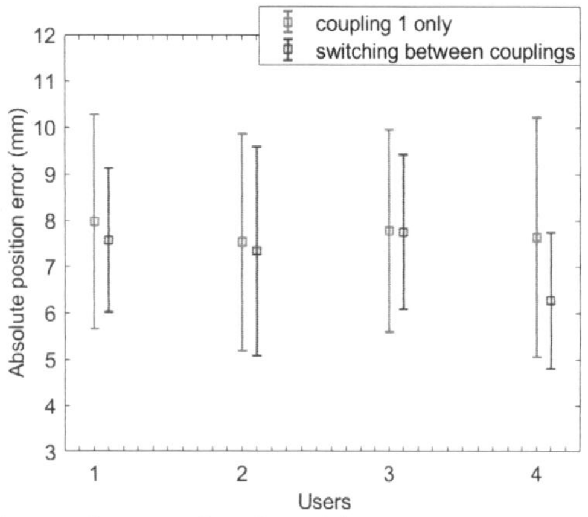

Figure 9. Bounds of position tracking errors of the remote manipulator with three unexperienced and one (4th) experienced users.

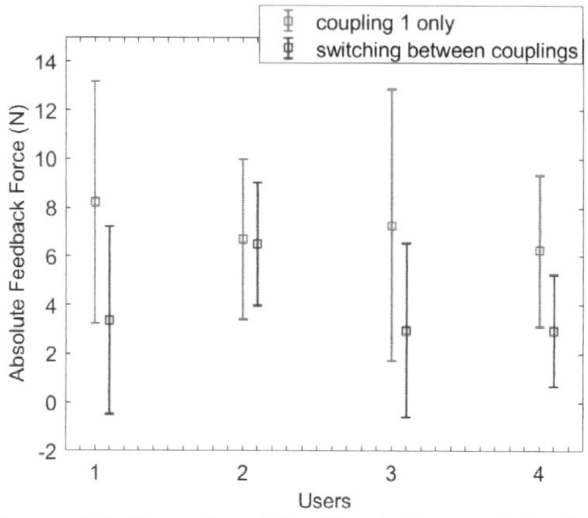

Figure 10. Bounds of feedback force of the local haptic device with three unexperienced and one (4th) experienced users.

Table 1 presents the total time spent by the users to complete the peg-in-a-hole operation. It is observed that switchings between couplings result with faster teleoperation with experienced and unexperienced users. The average operation time with unexperienced users is 81.3 s in the first case, and 66.5 s in the second case. The operation time with the experienced user are 66.4 s and 61.5 s respectively.

Table 1. Peg-in-a-hole operation time (in seconds).

Experiment	User 1	User 2	User 3	User 4
First Case (w/o haptic guidance)	99.20	70.29	74.30	66.38
Second Case (with haptic guidance)	66.66	67.57	65.23	61.46

Experimental results show on the one hand that the contribution of the haptic guidance, established by the switchings between couplings, is nearly negligible in position tracking performnace of the remote manipulator. On the other hand, a greater impact of the haptic guidance is observed in the feedback forces. In fact, the haptic guidance contributes to reducing the feedback forces, which in turn enables smoother physical interaction between the peg and hole. As a consequence, the total teleoperation time is reduced with the proposed haptic guidance by switching of the haptic couplings.

5. Conclusion

Design and implementation of the haptic teleoperation system for the peg-in-a-hole task is successfully achieved. In the experiments, stable operations are observed in presence of one experienced and multiple unexperienced users in the bilateral loop. Furthermore, contributions expected from the proposed controller design are also observed. Experimental results show that position tracking performance of the remote manipulator is improved with the introduction of the proposed haptic coupling. Besides, the interaction forces fed back to the user are attenuated and faster teleoperation of the peg-in-a-hole assembly are obtained independently from the user experience on the teleoperation system.

6. References

[1] Okamura A M 2009 *Current opinion in urology* **19**, p 102.
[2] Haidegger T, Sàndor J and Benyò Z 2011 *Surgical endoscopy*, **25**, p 681-690.
[3] Fan C, Zhang Y, Xie Z, Liu Y, Li Z and Liu H 2018 *IEEE International Conference on Mechatronics and Automation (ICMA)*, p 481-486.
[4] Desbats P, Geffard F, Piolain G and Coudray A 2006 *Industrial Robot: An International Journal*, **33** p 178-186.
[5] Liu X and Tavakoli 2018 *M Journal of Dynamics Systems, Measurement and Control*, **140**, p 121004.

[6] Nuño E, Basañez L, Lòpez-Franco C and Arana-Daniel A 2014 *Journal of the Franklin Institute*, **351**, p 241-258.
[7] Niemeyer G and Slotine J. J. 2018 *IEEE Journal of oceanic engineering*, **16**, p 304-313.
[8] Chopra N and Spong M W 2016 *Advances in Robot Control*, (A. Springer).
[9] Chopra N, Berestesky P and Spong M W 2008 *IEEE Transactions on Control Sytems Technology*, 16, p 304-313.
[10] Fu W, Van Paassen M R and Mulder M 2018 *IEEE International Conference on Systems, Man and Cybernetics*, p 4279-4284.
[11] Yang C, Wang X, Li Z, Li Y and Su C Y 2017 *IEEE Transactions on Systems, Man and Cybernetics: Systems*, **47**, p 2125-2136.
[12] Ganjefar S, Afshar M, Sarajchi M H and Shao Z 2018 *International Journal of Control, Automation and Systems*, **16**, p 2405-2420.
[13] Lu Z, Huang P and Liu Z. 2018 *IEEE Transactions on Industrial Electronics*, **65**, p 2439-2448.
[14] Bayraktaroglu Z Y, Argin O F and Haliyo S 2019 *Robotica*, **37**, p 338-357.
[15] van Oosterhout J, Wildenbeest J G, Boessenkool H, Heemskerk C J, de Baar M R, van der Helm F C and Abbink D A 2015 *IEEE Transactions on Haptics*, **8**, 164-175.
[16] Behandish M, Ilieş H T 2015 *Journal of Computing and Information Science in Engineering*, **15**, p041004.
[17] Bolopion A, Xie H, Haliyo D S and Règnier S 2012 *IEEE/ASME Transactions on Mechatronics*, **17**, p 116-127.
[18] Nudehi S S, Mukherjee R, Ghoudoussi M 2005 *IEEE Transactions on Control Systems Technology*, **13**, p 588-592.
[19] Stefanov N, Passenberg C, Peer A and Buss M 2013 *IEEE Transactions on Human-Machine Systems*, **43**, p 385-397.
[20] Basdogan C, Kiraz A, Bukusoglu I, Varol A, Doganay S 2007 *Optics Express*, **15**, p11616-11621.
[21] Colgate J E, Stanley M C and Brown J M 1995 *IEEE/RSJ In. Conf. Intell. Robot Syst.*, p 140-145.
[22] Stäubli RX series 160 family instruction manual, (2008).

Acoustic performance of locally resonant layer backing with panel-water-panel termination

Ye Han-Feng[1], Wang Shi-Peng[1], Wu Yue[1], Tao Meng[1]

[1]School of Mechanical Engineering, Guizhou University, Guiyang 550025, China
[1]Corresponding author: Tao Meng, Email: tomn_in@163.com

Abstract: The acoustic performance of locally resonant layer backing with the "panel-water-panel" termination has been studied. Using the finite element software COMSOL, the acoustic model of locally resonant layer has been given firstly. Then, the effect of several parameters including incident angles, cell numbers and cavity defects, on the sound absorption performance of locally resonant layer have been numerically analyzed.

Keywords: Acoustic Performance; Locally resonant layer; Oblique incidence; Cavity defect

1. Introduction

Locally resonant acoustic metamaterials are based on artificial periodic materials from sonic crystals. Usually, some cells which can be solid or liquid are distributed periodically in sonic crystals, so the parameters of different cells such as elastic constant and mass density are periodically alternated. In 1992, Sigalas [1] used spherical scatters to embed in a matrix to form a three-dimensional periodic lattice structure, and confirmed the existence of elastic band gaps for the first time theoretically. In 2013, Hwan [2] studied the attenuation and dissipation of acoustic waves in viscoelastic acoustic metamaterials which are composed of air and metal units. Based on the acoustic metamaterial which contains periodic criss-crossed elliptical holes, the band gap can be moved through altering the long and the short radii of elliptical hole by Gao [3]. Based on the extended transfer matrix method, Ansari [4] improved the wave propagation characteristics of sonic crystals by altering the band gap of the magnetoelastic material. Lü [5] studied the low-frequency sound absorption performance of the viscoelastic coating with different embedded scatterers.

At present, researches about locally resonant materials mostly focused on the infinite periodic structure, and the effect of material and structural parameters of cells on the band gap has been discussed mainly. But, there are not many studies concerning locally resonant materials used for the underwater sound absorption and noise reduction. In practical applications, it is impossible to get an locally resonant layer with an infinite thickness, and the acoustic performance of the locally resonant layer in a wider frequency range should be concerned, which was not covered in previous studies. Thus, the acoustic performance of a two-dimensional three-component locally resonant layer with a finite length under oblique incidence condition of plane wave is studied by the finite element software COMSOL in the next section.

2. Locally resonant layer

Figure 1 shows a two-dimensional three-component periodic locally resonant cell. As shown in the figure, component1, 2 and 3 are the coating made of silicon rubber, the oscillator made of lead and the matrix made of epoxy resin respectively. The densities are $\rho_i = 1300\ \text{kg/m}^3$,

$11600\,\text{kg/m}^3, 1180\,\text{kg/m}^3, (i=1,2,3)$, where subscript 1,2,3 indicates to silicon rubber, lead and epoxy resin respectively. Besides, the three Lame constants are $\lambda_i = 0.6$ MPa, 42.3 GPa, 4.43 GPa $(i=1,2,3)$, and the shear moduli are $\mu_i = 0.04$ MPa, 14.9 GPa, 1.59 GPa $(i=1,2,3)$, respectively. Moreover, the outer and inner radii (r1 and r2) of component1 are 6mm and 4mm respectively, and the length (H) of locally resonant cell is 14mm.The selection of materials of locally resonant layer is based on the Hou's research [6], but structural parameters have been changed in this paper. It should be noted that Hou focused on the infinite periodic structure mainly, while the finite one is taken as the study object in this paper.

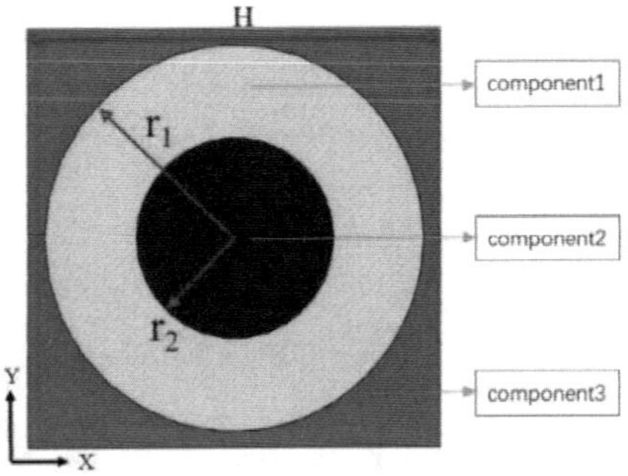

Figure 1. Schematic diagram of locally resonant cell

Figure 2 shows a model used to predict the underwater acoustic performance of a finite locally resonant layer whose thickness is 10 locally resonant cells. As shown in the figure, the perfect matching layer, the incident water region, the locally resonant layer, the steel panel 1, the intermediate water region and the steel panel 2 are fixed from left to right in the model. The thicknesses of steel panel 1, 2 and intermediate water are 6mm, 6mm and 180mm respectively.

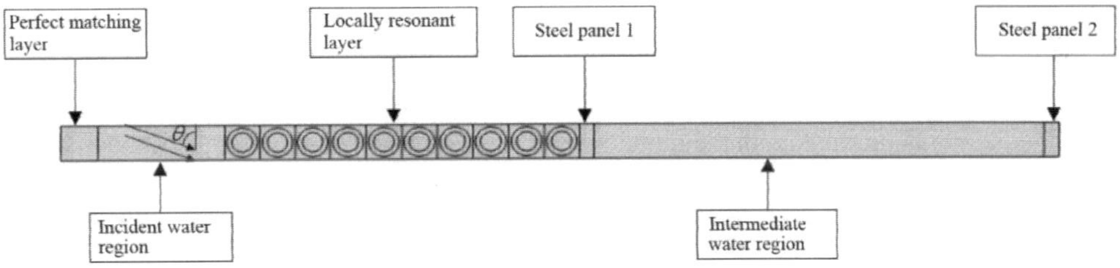

Figure 2. Prediction model of acoustic performance of locally resonant layer

3. Acoustic performance analysis of locally resonant layer

When the plane wave impinges normally on the locally resonant layer from the incident water, sound reflection coefficient and absorption coefficient of the locally resonant layer are shown in Figure 3. Because of the modulation of the intermediate water layer, the two curves of reflection coefficient and absorption coefficient show an obvious fluctuation with the increase of frequency.

Figure 4 compares the displacement fields of the locally resonant layer at the first peak frequency (600Hz) and the first trough frequency of absorption (2300Hz). As shown in the figure, the overall displacement of locally resonant layer at 600 Hz is obviously higher than the one at 2300 Hz: (1) at 600 Hz, vertical vibration of each resonant cell from top (the first one) to bottom (the tenth one) can be distinctly observed, and the maximum displacement whose value is up to 1.68×10-7m happens around the oscillator of the first resonant cell; (2) at 2300 Hz, the relatively obvious transverse vibration only can be found around the upper five resonant cells, and the displacement amplitude decreases significantly. Thus, if the displacement amplitude is higher and the numbers of vibrating cells is more, the friction between the oscillator, coating and matrix will be intenser, which will cause more loss of acoustic energy.

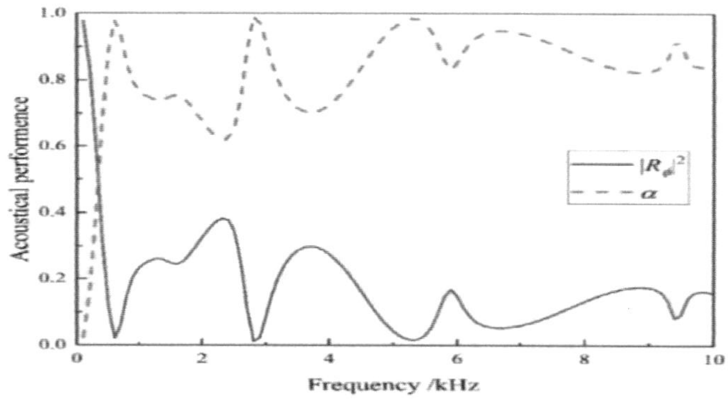

Figure 3. Acoustic performance of the locally resonant layer

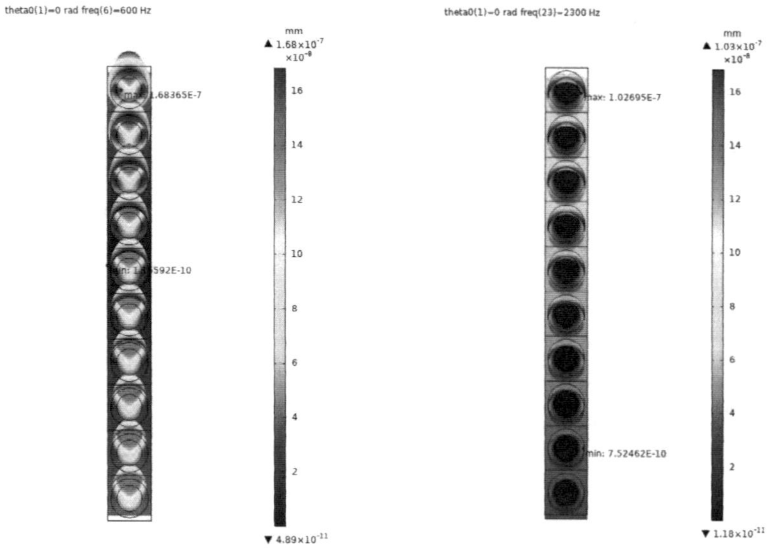

Figure.4 Displacement fields at 600Hz (left) and 2300Hz (right)

3.1. Effect of incident angle

Figure 5 compares the sound absorption coefficients of locally resonant layer under at different incident angles of plane waves. When the frequency is below 1000Hz, the change of incident angle has little influence on the sound absorption performance . With the increase of frequency, the sound absorption performance under normal incidence is better than those under other incident angles. Besides, the frequency of sound absorption coefficient at the peak will shift to a higher frequency if the incident angle increases.

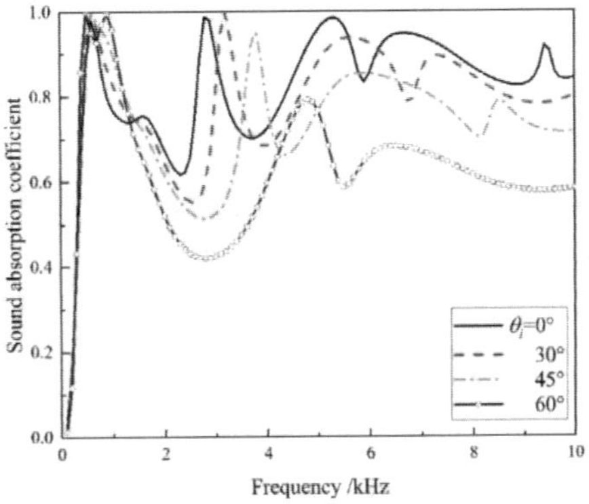

Figure 5. Effect of different incident angles on sound absorption coefficient of locally resonant layer

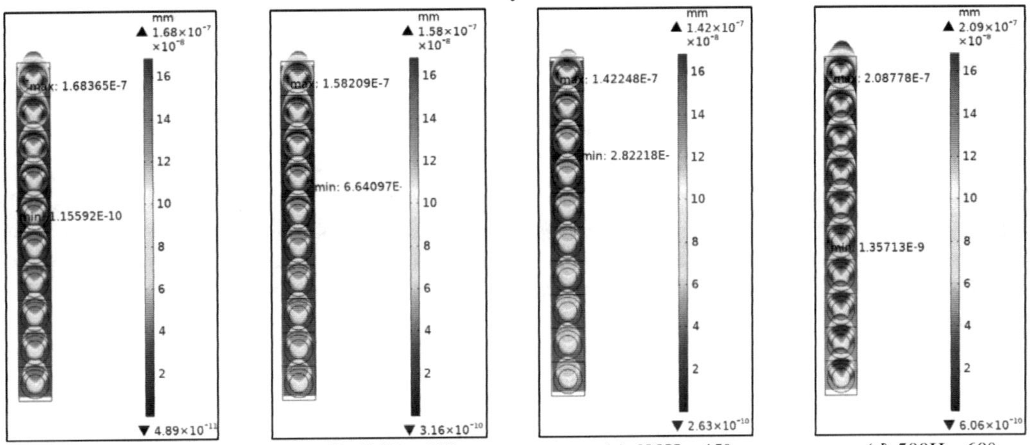

(a) 600Hz, 0° (*b*) 600Hz, 30° (*c*) 600Hz, 45° (*d*) 500Hz, 60°

Figure 6. Displacement field at the first absorption peak frequencies under different incident angles

Figure 6 shows the displacement fields of locally resonant layer at the first absorption peak frequencies (about 600, 600, 600 and 500Hz, because the frequency step is 100Hz in the acoustic simulation) under corresponding different incident angles (0°, 30°, 45° and 60°). As shown in the figure, the displacement fields of locally resonant layer are basically same under four different incident angles, only the location and the amplitude of the maximum and the minimum displacements are slightly different.

Figure 7 shows the energy loss of locally resonant layer at the first absorption peak frequencies under the corresponding different incident angles. It can be found that the energy loss is dominantly concentrated in the silicon rubber because of the vertical vibration of the oscillator. Since the incident energy will reflect mostly on the bottom surface of locally resonant layer because of the adjacent steel panel, and the more energy will accumulate and lose around the bottom of the locally resonant layer, so the maximum energy losses are all near the bottom oscillator at four incident angles.

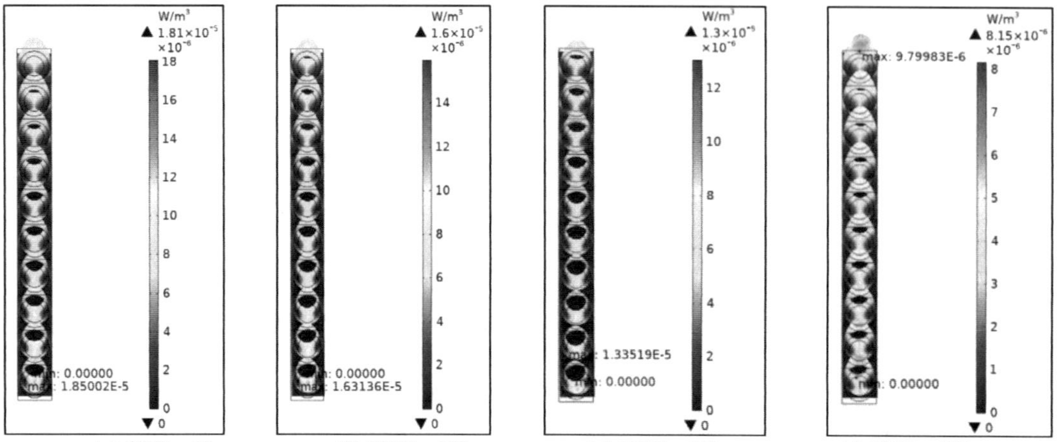

(*a*) 600Hz, 0° (*b*) 600Hz, 30° (*c*) 600Hz, 45° (*d*) 500Hz, 60°

Figure 7. Energy loss at the first absorption peak frequencies under different incident angles

3.2. Effect of cell number

The effect of the cell number (N = 1, 5, 10 and 15) on the acoustic performance of a locally resonant layer is shown in Figure 8, where other material properties and structural parameters remain unchanged. With the increase of cell number, the sound absorption performance of locally resonant layer improves greatly in the whole frequency range, especially in the low and the medium frequency ranges, and the frequency at the absorption peak shifts to a lower frequency.The sound absorption coefficients of 15 cells are over 0.75 in the whole frequency range, which is far more than the coefficients of 1 cell.

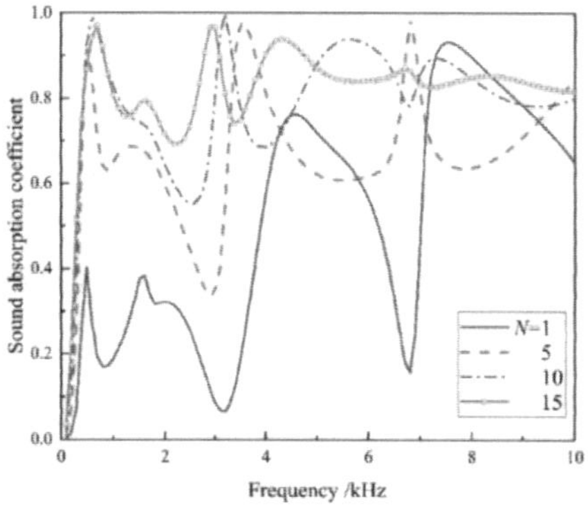

Figure 8. The effect of the number of cells on the acoustic performance

Figure 9 shows displacement fields of the locally resonant layer with different cell numbers at the second absorption peak frequency. As shown in Figure 9(c) and 9(d), the resonance of the whole layer mainly concentrates in the first five cells (from the top). When there is only one cell in the locally resonant layer, the energy loss is relatively lower, which leads to a smaller sound absorption coefficient. When cell number exceeds five, although the cell at the bottom of locally resonant layer vibrates as well, the amplitude is obviously weaker than the situation that cell number is 5, and the energy loss is very limited. Therefore, the improvement of sound absorption coefficient is very limited when the cell number increases to more than 5.

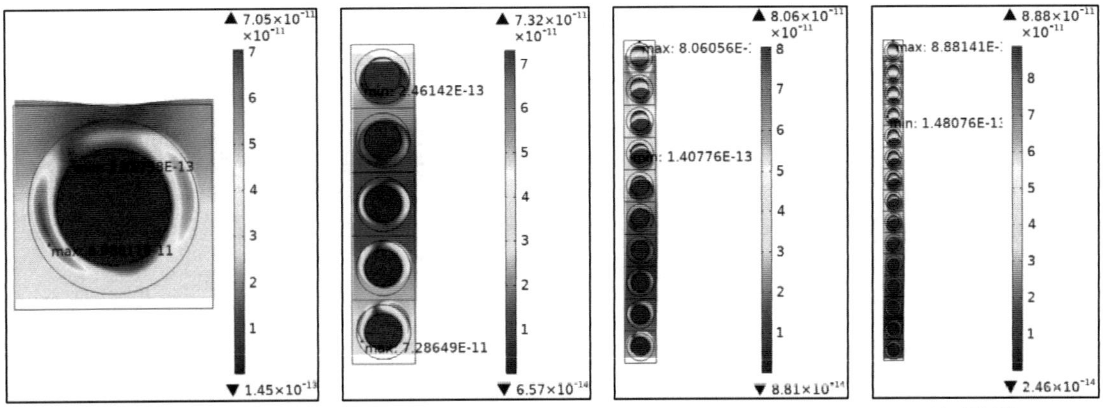

(*a*) 1500Hz (*b*) 1400Hz (*c*) 1400Hz (*d*) 1500Hz

Figure 9. Displacement field at the second absorption peak frequencies

3.3. Effect of cavity defect

It has been confirmed that the bandgap characteristics will be changed if some defects occur in an infinite periodic structure [7-8]. Taking account of 10-cell locally resonant layer shown in Figure 2, the lead oscillator of the first and fifth cells from the top (the incident side) are replaced by air to form two different locally resonant layers with cavity defects. Figure 10 shows the effect of cavity defects on the acoustic performance of locally resonant layer under the incident angle 30 degree. In this figure, the sound absorption coefficient of locally resonant layer which contains 10 alternately ordered oscillator units (lead-air-lead-air) is shown as green dotted lines.

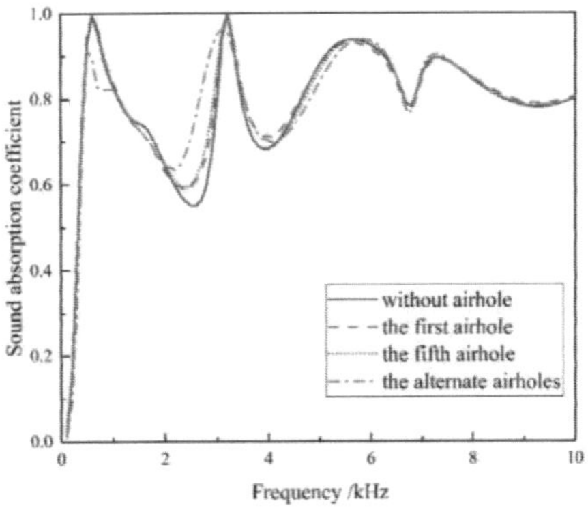

Figure 10. The effect of cavity defect on the acoustic performance

As shown in the figure, the sound absorption performance in the frequency range from 2100Hz to 3000Hz gets improved due to the existence of cavity defects. Compared with other defect forms,the layer in the alternate airholes defect form gets the best improvement of sound absorption from 2100Hz to3400Hz. In the frequency range from 3400Hz to 4600Hz, only the sound absorption performance of locally resonant layer with the first cavity defects has improved, but the range of improvement is not very wide. It can be concluded that the sound absorption performance of locally resonant layer in some frequency range will get improved if the oscillator changes from heavy lead to air.

4. Conclusion

Based on the numerical model of locally resonant layer, the acoustic performance of a two-dimensional three-component locally resonant layer with finite thickness has been analyzed, and some results are concluded as follows:

- For the plane in band-gap frequency range, the locally resonant layer with finite thickness can not completely block the wave propagating, but still can isolate most of the wave energy.
- As the cell number increases, the sound absorption performance of locally resonant layer gets improved effectively in the whole frequency range, especially in the low and the medium frequency ranges, and the frequency at the absorption peak shifts to a lower frequency.
- For finite locally resonant layer, the cavity defect can alternate the resonance characteristic and affect the acoustic performance of locally resonant layer in some frequency range and the range of improvement is not very wide.

5. References

[1] Sigalas M M, Economou E N. Elastic and acoustic wave band structure [J]. Journal of Sound &Vibration, 1992, 158(2): 377-382

[2] Hwan O J, Jae K Y, Young K Y. Wave attenuation and dissipation mechanisms in viscoelastic phononic crystals [J]. Journal of Applied Physics, 2013, 113(10): 212-301

[3] Gao N, Huang Y L, Bao R H, et al. Robustly tuning bandgaps in two-dimensional soft phononic crystals with criss-crossed elliptical holes [J]. Acta Mechanica Solida Sinica, 2018, 31(5): 573-588

[4] Ansari M H, Attarzadeh M A, Nouh M, et al. Application of magnetoelastic materials in spatiotemporally modulated phononic crystals for nonreciprocal wave propagation [J]. Smart Materials and Structures, 2017, 27(1): 015030

[5] Lǚ Lin-Mei, Wen Ji-Hong, Zhao Hong-Gang, et al. Low-frequency acoustic absorption of viscoelastic coating with various shapes of scatterers [J]. Acta Physica Sinica, 2012, 61(21): 514-518

[6] Lina H, Zhilin H, Xiujun F. Defect state of the locally resonant phononic crystal[J]. Acta Physica Sinica, 2014, 63(3):34305

[7] Benchabane S, Gaiffe O, Salut R, et al. Guidance of surface waves in a micron-scale phononic crystal line-defect waveguide [J]. Applied Physics Letters, 2015, 106(8):081903

[8] Gu C, Jin F. Research on the tunability of point defect modes in a two-dimensional magneto-elastic phononic crystal [J]. Journal of Physics D: Applied Physics, 2016, 49(17):175103

6. Acknowledgement

This work was supported by the National Natural Science Foundation of China (Nos. 51765008, 51365007), the High-level Innovative Talents Project of Guizhou Province (No. 20164033), and the Science and Technology Project of Guizhou Province (No. 20175781).

The in-plane dynamic crushing of quadrilateral chiral honeycomb coatings

Li Jun-Jie[1], Wu Yue[1], Wang Shi-Peng[1], Tao Meng[1]

[1]School of Mechanical Engineering, Guizhou University, Guiyang 550025, China

[1]Corresponding author: Tao Meng, Email: tomn_in@163.com

Abstract. For the anti-impact performance of hyperelastic coatings covered on the ship hull, the in-plane dynamic crushing behaviors of quadrilateral chiral honeycomb coatings (including chiral and anti-chiral) were numerically studied by explicit dynamic finite element simulations. The deformation mode, energy absorption and supporting reaction force under different impact speeds have been analyzed. The results show that: (1) If the impact speed is low, the chiral honeycomb coatings display the rotational deformation; (2) With the increase of impact speed, the deformation at the top of coating is remarkable; (3) The response of energy absorption and supporting reaction force corresponds to the macroscopic deformation mode of chiral honeycomb coatings.

Keywords: chiral honeycomb; hyperelastic rubber claddings; deformation characteristic; reaction force

1. Introduction

The coatings which are made of hyperelastic materials such as rubber are usually covered on the ship hull, of which the aim is to block the impact wave or isolate the sound propagation. In order to improve the performance of sound insulation and impact resistance, Some hyperelastic-viscoelastic coatings which are designed to include different kinds of holes or embedded scatterers have already been launched, especially some designed coatings which have the negative Poisson's ratio property attract more and more attentions.

As to now, lots of theoretical, numerical and experimental analyses of dynamic response about hyperelastic-viscoelastic coatings have been carried out [1-10]. Compared with the traditional coatings, the rotating deformation and the concave deformation, which belong to the chiral honeycomb structure with negative Poisson's ratio property, result in the better performance on the impact resistance and shear resistance. Lots of achievements are focused on the in-plane dynamics of porous honeycomb structures such as chiral honeycomb, but most of them are concentrated on thin-walled structures using ideal elastic-plastic materials to be the matrix. It has been confirmed that the chiral honeycomb coating which is made of hyperelastic materials can eliminate acoustic radiation effectively [4], also have the potential of impact resistance.

Taking the quadrilateral chiral and the quadrilateral anti-chiral honeycomb coatings as examples, the deformation mode, the energy absorption and the supporting reaction response of honeycomb coatings under the condition of in-plane impact are studied in the next sections, which is aimed to give the relationship between the macroscopic mechanical properties or the anti-impact performance with the geometric topology of the chiral structure.

2. Finite element model

Figure 1 shows the structural parameters of the quadrilateral chiral and the quadrilateral anti-chiral honeycomb cells, where *r*, *t*, and *l* are the outer radius of annulus, the thickness of ligament, and the center distance between two adjacent annulus respectively. First, a set of parameters including r = 3.4mm, t = 2mm, l = 9mm is used to shape the quadrilateral chiral honeycomb cell (C1) and the quadrilateral anti-chiral honeycomb cell (AC). Since the cross-sectional area of anti-chiral honeycomb cell is usually smaller than that of chiral honeycomb cell because of different arrangements, a new quadrilateral chiral honeycomb cell (C2) of which the cross-sectional area is the same as the anti-chiral honeycomb cell is also shaped.

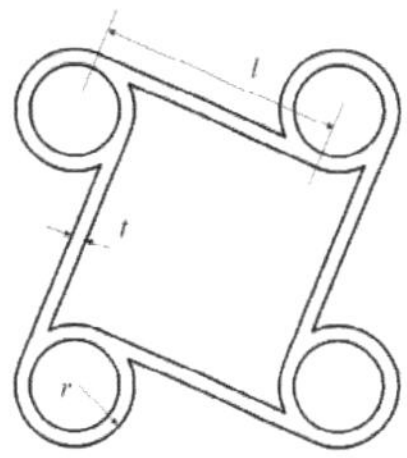

(a) a quadrilateral chiral cell

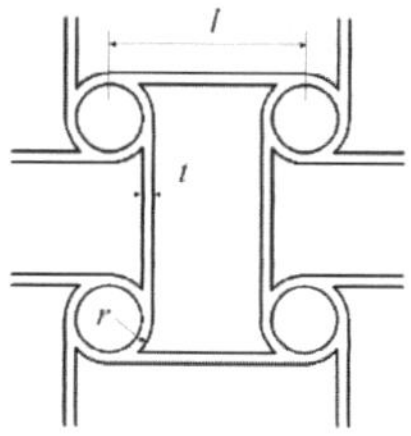

(b) a quadrilateral anti-chiral cell

Figure. 1. Schematic diagrams of quadrilateral chiral and anti-chiral honeycomb cells

The relative density of the honeycomb coating which contains quadrilateral chiral holes can be given as follows:

$$\bar{\rho} = \frac{\rho^*}{\rho_s} = \sum_{i=1}^{n} \frac{l_i t_i}{L_1 L_2} \tag{1}$$

Where the density of the honeycomb coating is ρ^*, the density of matrix is ρ_s, the ligament length and the corresponding thickness of the *i*th cell are l_i and t_i, and the overall length and height are L_1 and L_2 respectively.

According to Equation (1), some parameters of different honeycomb coatings are shown in Table 1.

Table 1. Parameters of different honeycomb coatings

Coating	r /mm	t /mm	l /mm	L_1/mm	L_2/mm	$\bar{\rho}$
Chiral honeycomb (C1)	3.5	2.0	9.00	103.5	62.5	0.539
Chiral honeycomb (C2)	3.5	2.0	7.48	92.0	56.0	0.599
Anti-chiral honeycomb (AC)	3.5	2.0	9.00	92.0	56.0	0.663

The dynamics software ABAQUS is used to analyze the impact response of honeycomb coatings which contain chiral or anti-chiral holes numerically. Figure 2 shows the two-dimensional model in the software. The rigid plate is meshed by the two-dimensional element of rigid joint, and the honeycomb coating is meshed by the stress element of reduced integral plane. In order to confirm the convergence of stress and strain at the corner of cell and other large deformation regions, the size of grid is assumed to be 0.5mm. The relationship between the rigid plate and the honeycomb coating at their interface is described by a face-to-face contact algorithm. To prevent contact penetration of the matrix during compression deformation, self-contact is adopted inside the honeycomb coating. Compared to the ship hull, the stiffness of honeycomb coating is usually much lower, so the bottom of honeycomb coating is set to be fixed if the impact wave is normally incident (or along the y direction), which is the same as the boundary conditions given by Zhang [6]. But, the left and the right boundary conditions should be symmetric in order to describe the horizontal periodicity of honeycomb coating according to Xiao [3]. The effective anti-impact performance of honeycomb coating requires that the coating should not be completely densified when compressed, so the nominal strain $\varepsilon = 0.6$ is taken as the nominal amount of compression in the y direction.

Figure 3 shows some nominal stress-strain data of rubber measured by the uniaxial, the biaxial and the plane experiments [11]. The constitutive model Ogden_N2 which can accurately describe the nonlinear relationship between stress and strain in the large strain interval, is used to fit the material parameters as follows: μ_1=0.375810163MPa, λ_1=1.88384694MPa, μ_2=0.001140191MPa, λ_2=-2.86492799MPa, D=0, and the density of rubber is 1500kg/m^3.

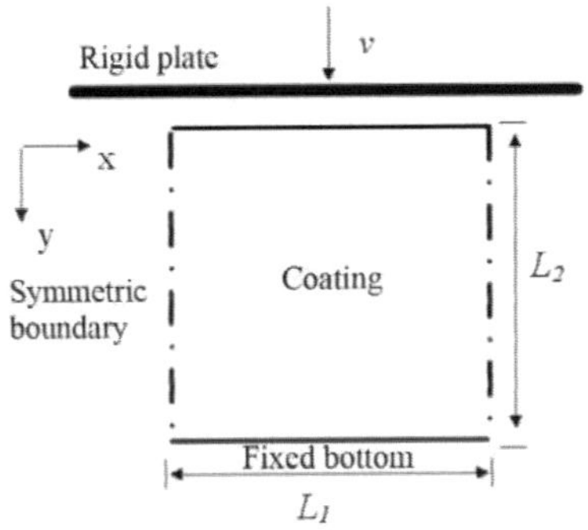

Figure 2. Diagram of honeycomb coating under in-plane impact.

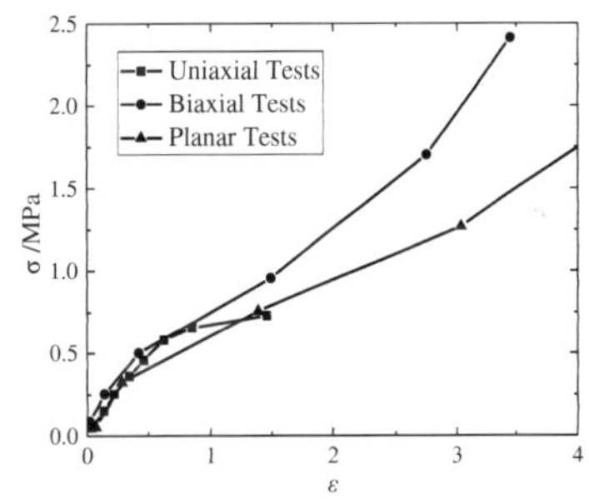

Figure 3. Tested data of rubber material

3. Dynamic responses of chiral honeycomb coatings

Figure 4 shows the typical deformation mode of the quadrilateral chiral honeycomb coating (C1) under different impact velocities. Under the quasi-static velocity 1m/s, the compression stress wave is gradually transferred from the top to the bottom of coating, and the weak deformation of cells can be observed. After the compression stress wave reaches the bottom of coating, the wave will be reflected from the bottom to the top reversely. Thus, the stress wave is transmitted up and down through the annulus and the ligament of cells, and all circular holes are gradually compressed into elliptical holes. Correspondingly, the ligaments of cells are rotated when the circular holes are under compressed deformation.

With the impact velocity increased to 5m/s, the circular hole appears a layered buckling deformation from the top to the bottom, but it is not compacted. During the compressed process, the ligaments and the annuli of cells are rotated obviously. Moreover, with the impact velocity increased higher to 10m/s, the inertia effect is significant, so the circular hole wall at the top of coating is crushed instantaneously. With the downward compression, the walls of circular holes are crushed layer by layer, and the response at the bottom is obviously delayed compared to that at the top of honeycomb coating.

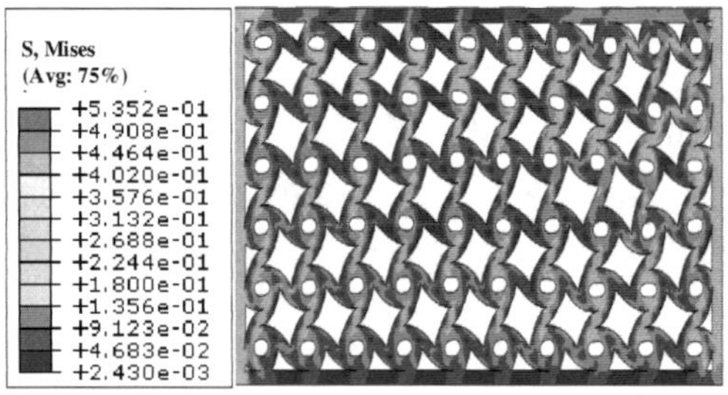

(a) v=1m/s

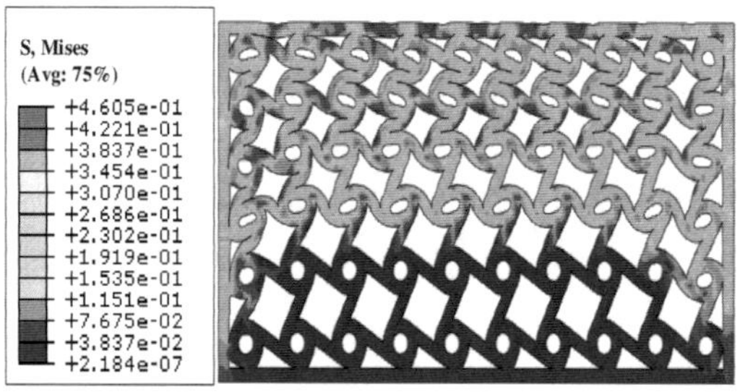

(b) v=5m/s

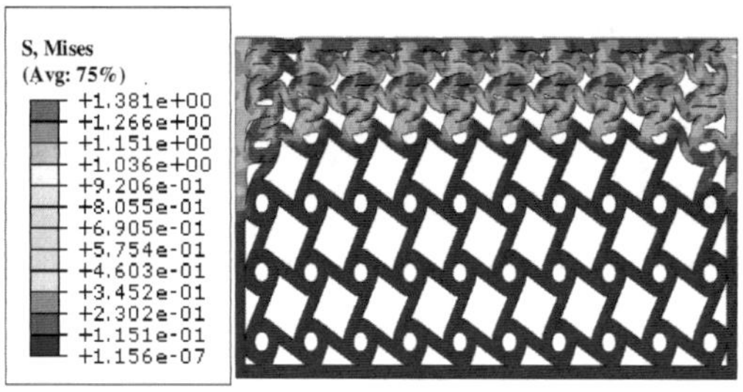

(c) v=10m/s

Figure 4. Deformation mode of chiral honeycomb coating (C1) under different impact velocities

Figure 5 shows the typical deformation mode of the quadrilateral anti-chiral honeycomb coating (AC) under different impact velocities. Under the quasi-static velocity 1m/s, the compressive stress wave is transmitted layer by layer from the top to the bottom, and the cell is little deformed. With further compression, the longitudinal and the transverse cell walls come into contact.

With the impact velocity increased to 5m/s, the upper circular holes appear buckling deformation firstly, and the circular holes present torsional deformation according to the clockwise and the counterclockwise directions alternately, which is due to the arrangement of anti-chiral honeycomb coating. Furthermore, with the impact velocity increased higher to 10m/s, it is similar to the chiral honeycomb coating that the top of coating is significantly deformed, which is consistent with the I-shaped compaction collapsed of the chiral honeycomb coating under high-speed impact.

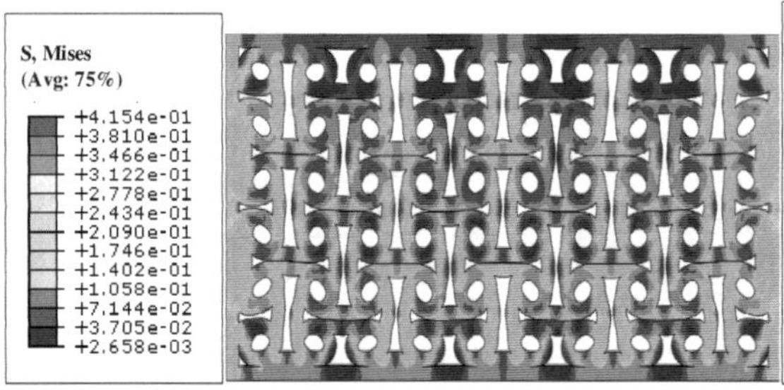

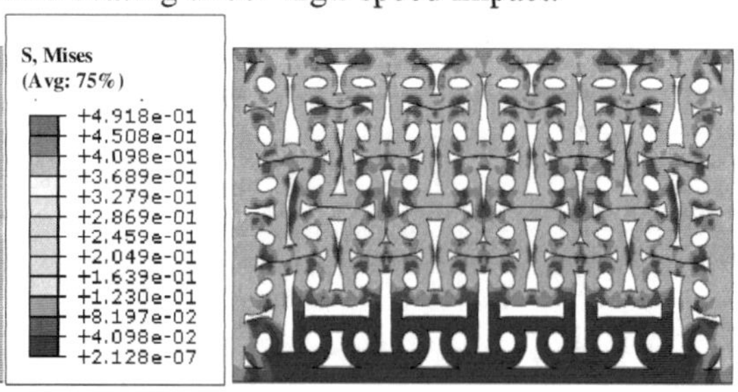

(*a*) *v*=1m/s (*b*) *v*=5m/s

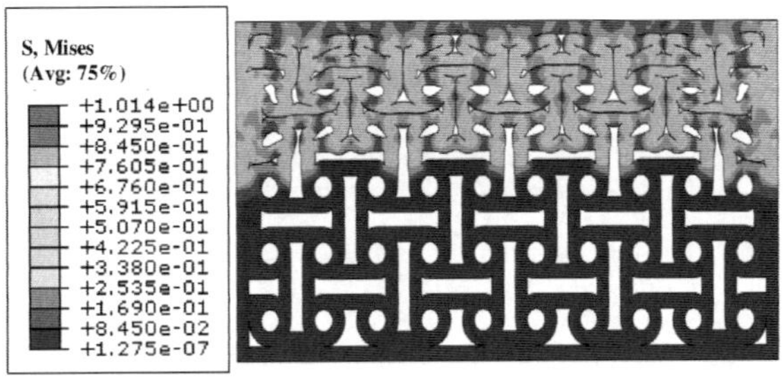

(*c*) *v*=10m/s

Figure 5. Deformation mode of anti-chiral honeycomb coating (AC) under different impact velocities

4. Energy absorption of chiral honeycomb coatings

The kinetic energy and the structural deformation energy resulting from the compressive motion of chiral honeycomb coating are the most important energy absorption forms. The energy absorption rate (SEA) is an important parameter to estimate the energy absorption capacity of multicellular structures [12]:

$$E_m = \frac{W_v}{\bar{\rho}\rho_s} \tag{2}$$

Where W_V is the absorbed energy per unit volume, $\bar{\rho}$ is the relative density of coating, and ρS is the matrix density.

According to Equation (2), Figure 6 shows the energy absorption rate (SEA) of three different chiral honeycomb coatings under different impact velocities. The energy absorption capacity of coating becomes better with the increase of impact velocity, especially before the whole coating is shifted into compaction stage. In addition, the anti-chiral honeycomb coating (AC) exhibited much stronger energy absorption capacity than the chiral honeycomb coating (C1) under the assumption of same cell parameters, and also than the chiral honeycomb coating (C2) under the assumption of same volume.

5. Supporting reaction force of chiral honeycomb coatings

Figure 7 shows the relationship between the supporting reaction force and the compressive nominal strain of three different chiral honeycomb coatings under different impact velocities. In the quasi-static process (v = 1m/s), the overall trend of supporting reaction force curves of three different coatings becomes higher with the increasing of nominal strain. The honeycomb hole wall appears linear elastic bending under the low impact velocity, and the linear relationship between the reacting force and strain can be approximately obtained. After that, the hole walls are buckled and collapsed, and contacted to each other. As a result, the overall stiffness of chiral honeycomb coating increases, and also the supporting reaction force increases obviously. The collapsed process of honeycomb coating layer by layer during the crashing compression causes the curve of supporting reaction force to oscillate.

When the impact velocity is v = 5m/s, the supporting reaction force of three different coatings is increased from zero when the nominal strain reaches around 0.2, and the increasing rate is faster than that under quasi-static condition. Then, the value of supporting reaction force reaches the first peak,

IOP Conf. Series: Materials Science and Engineering **707** (2019) 012017 doi:10.1088/1757-899X/707/1/012017

which is caused by the compressive stress wave transmitted from the top to the bottom of coating. Next, the honeycomb hole walls is collapsed layer by layer and the supporting reaction force decreases a little. After the compression goes to the compaction stage, the supporting reaction force reaches the second peak value, which is higher than the first peak value and the ascending rate is faster. Moreover, the peak value of supporting reaction force of the coating (AC) is the highest in the three different coatings in the whole process.

When the impact velocity is v = 10m/s, the supporting reaction force of three coatings is almost equal to zero in the early stage of impact response, and the bottom is not subjected to stress before the compression wave transmitted to the bottom of coating. When the strain reaches to 0.35 approximately, the reacting force is ascended rapidly. The ascending rate of reacting force of coating (AC) is faster than that of C1 and C2, and reaches the higher peak when the strain is smaller. This is directly related to the distribution of quadrilateral chiral cells: the cells of coating (AC) are distributed more closely than that of C1 under the assumption of same geometric parameters, or the relative density of coating (AC) is greater than that of C2 under assumption of same volume. In the compaction stage, the supporting reaction force curves of three coatings all oscillate up and down. Totally speaking, the response of supporting reaction force of chiral honeycomb coating varies greatly with the change of impact velocity, and there is a corresponding relationship between the response of supporting reaction force and the deformation process of coating.

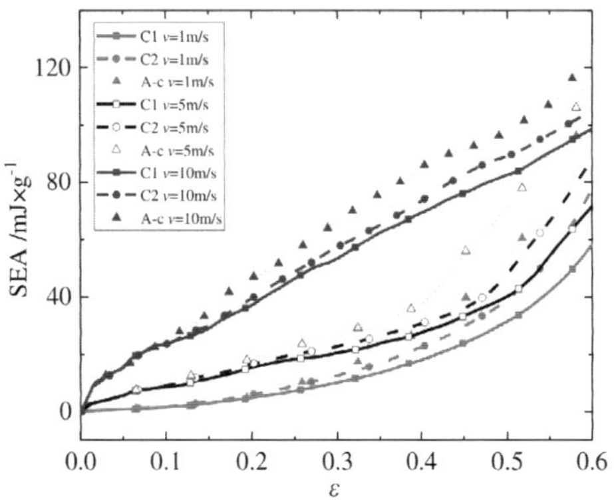

Figure 6. SEA of three chiral honeycomb coatings under different impact velocities

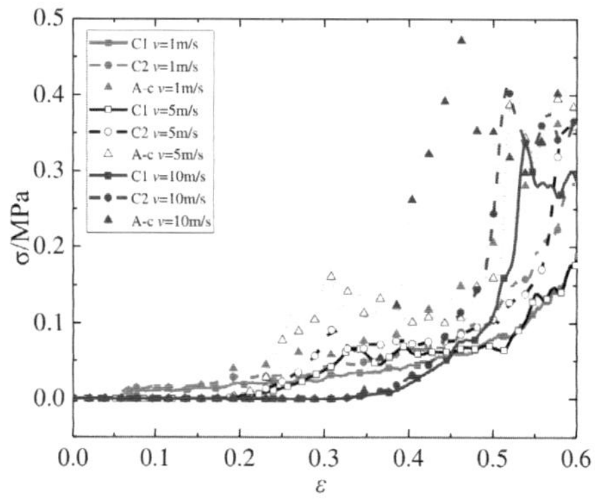

Figure 7. Reacting force of three chiral honeycomb coatings under different impact velocities

6. Conclusions

Based on the two-dimensional dynamic compression model, the in-plane impact dynamics responses of the quadrilateral chiral and the quadrilateral anti-chiral honeycomb coatings have been numerically analyzed. It is found that:

(1) Impact velocity affects the deformation modes of chiral honeycomb coatings. Under a lower velocity, the rotation of honeycomb cell and the ligaments twining with each other is the prominent deformation. With the increase of impact velocity, the deformation of the coating top is much significant.

(2) The performance of energy absorption by the coating becomes better with the increase of impact velocity or strain. Under the same impact velocity, the energy absorption of anti-chiral honeycomb coating is the highest.

(3) The response of supporting reaction force is related to the macroscopic deformation of coating. Under a lower impact velocity, the deformation is uniform, and the supporting reaction force is raised with the increase of strain. If the impact velocity increases, the supporting reaction force is greatly affected by compression stress and the coating deformation. The effective stiffness of quadrilateral chiral honeycomb coating is less than that of anti-chiral honeycomb coating, which results in the smaller supporting reaction force under the same nominal strain.

7. References

[1] Ruan D, Lu G, Wang B, et al. In-plane dynamic crushing of honeycombs—a finite element study J 2003 *International Journal of Impact Engineering*, **28**:161-182.

[2] Liu X N, Huang G L, Hu G k. Chiral effect in plane isotropic micropolar elasticity and its application to chiral lattices J 2012 *Journal of the Mechanics & Physics of Solids* **60** 1907-1921

[3] Alderson A, Alderson K L, Attard D, et al. Elastic constants of 3-, 4- and 6-connected chiral and anti-chiral honeycombs subject to uniaxial in-plane loading J 2010 *Composites Science & Technology*, **70**:1042-1048.

[4] Lorato A, Innocenti P, Scarpa F, et al. The transverse elastic properties of chiral honeycombs J. 2010 *Composites Science & Technology*, **70**:1057-1063.

[5] Reis F D, Ganghoffer J F. Equivalent mechanical properties of auxetic lattices from discrete Homogenization J. 2012, *Computational Materials Science*, **51**:314-321.

[6] Zhang X C, Liu Y, Li N. The In-plane dynamic crushing of honeycombs with negative Poisson's effects J 2012 *Explosion and Shock Waves* **32** 475-482

[7] Xiao F, Chen Y, Sun J Y, et al. Dynamic crush behavior and performance of layered honeycomb hyperelastic rubber claddings *J* 2013 *Journal of Vibration and Shock* **32** 13-20

[8] Gao Q, Wang L, Zhou Z, et al. Theoretical, numerical and experimental analysis of three-dimensional double-V honeycomb J 2018 *Materials & Design* **139** 380-391

[9] Zou T, Zhou L, Mechanical property analysis and experimental demonstration of zero Poisson's ratio mixed cruciform honeycomb J 2017 *Materials Research Express* **4** 4 045702

[10] Lim, Teik-Cheng. A 3D auxetic material based on intersecting double arrowheads J 2016 *physica status solidi (b)* **253** 1252-1260

[11] ZHUANG Zhuo, ZHANG Fan, CEN Song, et al. The nonlinear finite element analysis and example of ABAQUS M. 2005. *Beijing: Science Press*

[12] Kooistra G W, Deshpande V S, Wadley H N G. Compressive behavior of age hardenable tetrahedral lattice truss structures made from aluminium J. 2004 *Acta Materialia,* **52**:4229-4237.

8. Acknowledgements

This work was supported by the National Natural Science Foundation of China (Nos. 51765008, 51365007), the High-level Innovative Talents Project of Guizhou Province (No. 20164033), and the Science and Technology Project of Guizhou Province (No. 20175781).

IOP Conf. Series: Materials Science and Engineering **707** (2019) 012018 doi:10.1088/1757-899X/707/1/012018

Adaptive stabilization of two complex-conjugate dominant poles and quality indices of oscillating transient process in systems with interval parametric uncertainty

I V Khozhaev[1]

[1] School of Computer Science and Robotics, National Research Tomsk Polytechnic University, Tomsk, 634034, Russia

Abstract. The paper is dedicated to a development of a controller synthesis method for control systems with an interval parametric uncertainty. The method combines two approaches to controller synthesis: adaptive one and interval one; also it is based on a pole domination principle. Therefore, the method allows providing desired control quality by placing two complex-conjugate dominant poles as necessary and reducing an influence of all other poles on a transient process in conditions of an interval parametric uncertainty. Application example of the synthesis method is provided; results of simulation modelling of a system, synthesized with the help of the method, are also provided.

1. Introduction

A vast variety of control objects or systems have uncertain parameters. Such parametric uncertainty must be considered during mathematical modelling or synthesizing control systems, which are to be reliable and robust, for example: industrial robots, aerial or underwater unmanned vehicles, etc.

There are several common approaches to synthesizing controllers for control systems with parametric uncertainty. The most widely used of them are adaptive control, fuzzy-logic control, neural-networks-based control and robust control, which considers uncertain parameters to vary in some known intervals. Each of these approaches has its own advantages and disadvantages. Adaptive control systems have stable control quality, but cannot adapt to rapid wide-range variations of parameters [1-3]. Neural-network-based controllers are able to provide desired control quality in such conditions, but require a vast amount of data to train a neural network [4]. Also, both adaptive controllers and neural-network based controllers have a complex structure. Fuzzy-logic controllers have a simpler structure, but the procedure of synthesizing a base of rules is not formal and requires a highly-qualified engineer to perform it [5]. Robust controllers have simple structure, though they are common linear controllers with constant parameters; procedures of their synthesis are formalized and require minimum data: only minimal and maximal estimates of uncertain parameters, but robust control systems often have insufficient dynamics and value of their quality indices may vary slightly due to parametric variations [6-14]. So, the problem of developing new methods of synthesizing a controller is highly relevant.

One of ways to improve existing synthesis methods is to combine them. This paper is dedicated to development of combined method of controller synthesis, which is based on adaptive and robust control simultaneously. The method will have advantages of both of these approaches: desired system dynamics, stable values of control quality indices, an ability to cope with rapid wide-range parametric variations, a minimum amount of data required, a simple structure of a controller and a strictly formalized synthesis algorithm. To develop such synthesis method is the main aim of the research. In order to reach the aim, a set of objectives must be accomplished: the problem must be set in terms of mathematics, general expression for controller synthesis must be derived, an algorithm of an adaptive-robust controller synthesis must be formulated and the algorithm must be applied to a synthesis problem. Let us now consider accomplishing each of these objectives.

2. Formulation of the problem

The method is based on the robust approach to control, which considers all uncertain parameters of a system to vary in known limits or intervals. Each control system can be described by its characteristic polynomial, which is a denominator of a transfer function of a closed-loop system. Coefficients of characteristic polynomial include all interval parameters of the system and, consequently, may also vary in some intervals. Let us designate a characteristic polynomial with interval coefficients as an interval characteristic polynomial (ICP). General form of an ICP can be written as follows:

$$D(s,\vec{K}) = \sum_{i=0}^{n}\left[q(\vec{K})\right]_i \cdot s^i = \sum_{i=0}^{n}\left[\underline{q_i};\overline{q_i}\right]\cdot s^i, \quad (1)$$

where $D(s,\vec{K})$ – is an ICP, $\vec{K}$ – is a vector of controller parameters, $q_i(\vec{K})$ – is an ICP coefficient function of parameters of a controller, $\underline{q_i}, \overline{q_i}$ – are minimal and maximal values of each ICP coefficient, which include parameters of a controller.

Roots of (1), which are also poles of a system, migrate inside some regions of a complex plane due to parametric variations. This leads to instability of control quality indices: stability degree and oscillability degree. In order to reduce variations of control quality, a pole domination principle can be applied. According to this principle, in order to provide desired control quality in a system, it is enough to place one or two dominant poles according to desired values of stability degree and oscillability degree and then place all other non-dominant poles far away of dominant ones to reduce their influence on a transient process [15-17]. There are two practically significant ways to place dominant poles on a complex plane: one real dominant pole and two complex-conjugate dominant poles. The first variant allows providing an aperiodic transient process of desired setting time; the second variant allows determining a stability degree and an oscillability degree simultaneously. In the paper the second variant is considered: placing two complex-conjugate dominant poles. An example of such allocation is given in the figure 1.

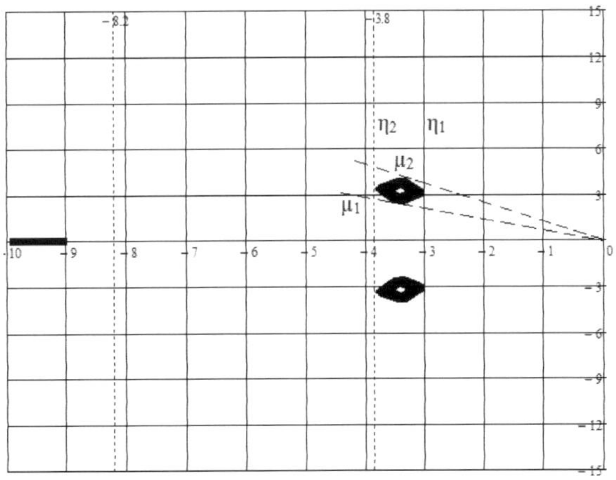

Figure 1. Example of poles allocation of an interval control system with two complex-conjugate dominant poles.

In the figure 1 complex-conjugate dominant poles lie provide following values of quality indices: stability degree varies from $\eta_1 = 3$ to $\eta_2 = 3.8$; oscillability degree varies from $\mu_1 = 0.7$ to $\mu_2 = 1$. All other poles lie on the right of the line $\mathrm{Re}(x) = -8.2$ and their influence on a transient process is minimized. The aim of the paper is to develop a method, which will allow stabilizing dominant poles in points of a complex plane, but not regions of it. This will make control quality of synthesized systems more stable. Let us now consider the main idea of the method.

The main idea of the method consists in decomposing an ICP in three parts: a dominant polynomial $A(s)$; a non-dominant polynomial $B(s,\vec{K})$, which is a result of ICP division by dominant polynomial $A(s)$, and a remainder $R(s,\vec{K})$ of such division. Decomposed ICP can be written as follows:

$$D(s,\vec{K}) = A(s)\cdot B(s,\vec{K}) + R(s,\vec{K}). \quad (2)$$

In the paper a problem of placing two dominant complex-conjugate poles is considered, so, dominant polynomial $A(s)$ can be written as follows:

$$A(s) = (s-\lambda_1)(s-\lambda_2) = (s-\alpha-j\beta)(s-\alpha+j\beta), \quad (3)$$

where λ_1, λ_2 – are dominant complex-conjugate poles; α, β – are their real and imaginary part.

Synthesis of a controller can be performed in three steps: defining dominant poles, placing roots of $B(s,\vec{K})$ far enough of dominant poles and providing $R(s,\vec{K})$ equality to zero. Now, we can formulate the problem of the research: to develop a method of placing roots of $B(s,\vec{K})$ and providing $R(s,\vec{K})$ equality to zero despite interval parametric uncertainty of both expressions.

3. Derivation of general expressions

Let us now derive expressions, which will allow calculating coefficients of a non-dominant polynomial and a remainder. As son, as we know the general form of an ICP (1) and a form of a dominant polynomial (3), we can divide (1) by (3) in general form and derive necessary expressions. They can be written as follows:

$$[b]_i = [q]_{i+2} + x\cdot[b]_{i+1} - y\cdot[b]_{i+2}, i \in n-2...0; \quad (4)$$

$$R(s) = (q_1(\vec{z}) + x\cdot b_0(\vec{z}) - y\cdot b_1(\vec{z}))\cdot s + q_0(\vec{z}) - y\cdot b_0(\vec{z}); \quad (5)$$

where $[b]_i$ – are coefficients of non-dominant polynomial; x – is a sum of dominant poles; y – is a production of dominant poles; n – is a degree of an ICP; $R(s)$ – is a remainder; $q_0(\vec{z}), q_1(\vec{z}), b_0(\vec{z}), b_1(\vec{z})$ – are dependencies between coefficients of an ICP and a dominant polynomial and a vector of interval parameters $\vec{z}$. Let us now formulate the synthesis algorithm on a base of (4) and (5).

4. Development of the synthesis algorithm

In order to formulate the synthesis algorithm we will define required data, algorithm itself and resulting data.

First of all, before starting the synthesis procedure, the aim of the synthesis should be defined: desired values of stability degree and oscillability degree. The method is based on a pole domination principle, so, these parameters will be defined through desired allocation of dominant poles of a system. Also, parameters of a non-dominant poles allocation area must be define: maximal real part of a non-dominant pole and sector angle of non-dominant poles allocation. All these data will define a desired control quality of a system; an example of such definition is given in the figure 2.

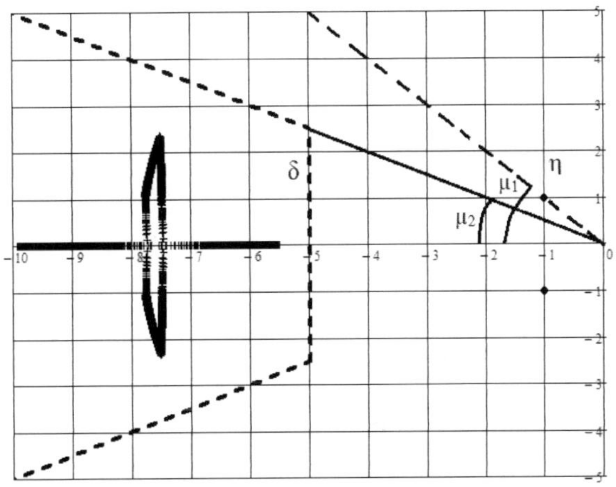

Figure 2. Example of defining a desired control quality through the system poles allocation

In the figure 2 robust stability degree of the system η and robust oscillability degree of the system μ_1 is defined by dominant poles. Also in the figure 2 a maximal real part of a non-dominant pole δ and sector angle of non-dominant poles allocation μ_2 are defined. It should be noticed, that μ_2 must not exceed μ_1.

Now, after the aim of synthesis is determined, a controller structure should be chosen. Transfer function of the controller must include three and more tunable parameters: two parameters are necessary to stabilize a pair of dominant poles; one parameter is necessary to place all non-dominant poles. In example of the method application a PID-controller will be used. Transfer function of a PID-controller can be written as follows:

$$W_{PI}(s, K_P, K_I, K_D) = K_P + \frac{K_I}{s} + K_D \cdot s, \tag{6}$$

where K_P, K_I, K_D – are proportional, integral and differential coefficients of PID-controller.

After choosing the controller, a transfer function of a whole system can be derived. Coefficients of denominator of such function, which is an ICP in a form (1), can be used in (4) and (5) to calculate coefficients of a non-dominant polynomial $B(s, K_P, K_I, K_D)$ and a remainder $R(s, K_P, K_I, K_D)$. After deriving a remainder, we should provide its equality to zero. To do this, we perform a substitution $s \to j\omega$, equate real and imaginary part of a remainder to zero and join these two equations in a system. In case of using a PID-controller (6) this system can be written as follows:

$$\begin{cases} \mathrm{Re}(R(j\omega, K_P, K_I, K_D, \vec{z})) = 0; \\ \mathrm{Im}(R(j\omega, K_P, K_I, K_D, \vec{z})) = 0. \end{cases}$$

By solving this system of equations functions, linking parameters of controller and interval parameters of a system with each other can be derived. General form of such function can be written as follows:

$$\begin{aligned} K_P &= f(K_D, \vec{z}); \\ K_I &= f(K_D, \vec{z}). \end{aligned} \tag{7}$$

These functions are laws of controller adaptation, which allow stabilizing dominant poles in points of a complex plane. After deriving, these function must be substituted to expressions of $B(s, K_P, K_I, K_D)$ coefficients. This will allow deriving $B(s, K_D)$.

After expression for non-dominant polynomial $B(s, K_D)$, which depends from only one parameter of a controller, is derived, a D-partition method can be used to place its roots in desired area of a complex plane. An equation of a D-partition curve can be derived by substituting s in $B(s, K_D)$ with a border equation of a non-dominant poles allocation area. For example, in order to provide a minimal real part of non-dominant poles, following substitution must be performed:

$$s \to j \cdot \omega + \delta, \delta \le 0, \tag{8}$$

In order to place non-dominant poles in a sector, a following substitution must be performed:

$$s \to \frac{1}{\mu_2} \cdot \omega + j \cdot \omega, \mu \ge 0, \tag{9}$$

Performing a D-partition with both substitutions will allows placing non-dominant poles in a truncated sector, as shown in the figure 2. Constant value of the differential coefficient K_D, chosen from a D-partition, will provide desired allocation of non-dominant poles.

Final form of the synthesized controller consists in laws of adaptation for proportional K_P and integral K_I coefficients (7) and a constant value of a differential coefficient K_D, chosen from a D-partition.

To sum up all aforementioned information, the synthesis algorithm can be formulated:

1. Choose desired control quality and provide all required date as shown in the figure 2.
2. Choose an appropriate controller; derive a transfer function of the system and find the ICP in a form (1).
3. Derive coefficients of the non-dominant polynomial $B(s,\vec{K})$ and the remainder $R(s,\vec{K})$ with the help of (4) and (5).
4. Find adaptation laws (7) for two parameters of the controller.
5. Substitute (7) to $B(s,\vec{K})$ and derive equations of D-partition curves with the help of substitutions (8) and (9).
6. Choose values of other parameters of the controller from D-partition.
7. Controller is synthesized. It consists of adaptation laws for two of controller parameters and constant values of the rest of them.

Let us consider an application example of the algorithm.

5. Example of synthesis method application

Let us assume, that there is a problem of providing a desired control quality in a control system with interval parametric uncertainty. Transfer function of a control object of the system is given below:

$$W(s) = \frac{1}{[p_3] \cdot s^3 + [p_2] \cdot s^2 + [p_1] \cdot s + [p_0]},$$

Where $[p_3]=[1;1.1]$, $[p_2]=[28;32]$, $[p_1]=[215;285]$ and $[p_0]=[350;370]$ – are interval parameters of the system.

According to the synthesis algorithm, let us now determine a desired control quality: $\eta = 1$, $\delta = -10$, $\mu_1 = \mu_2 = 1$. In order to reach the aim of synthesis a PID-controller (6) will be used. Considering this, an ICP of the system can be written as follows:

$$D(s, K_P, K_I, K_D, \vec{p}) = [p_3] \cdot s^4 + [p_2] \cdot s^3 + ([p_1] + K_D) \cdot s^2 + ([p_0] + K_P) \cdot s + K_I.$$

Let us now derive the non-dominant polynomial with the help of (4):

$$\begin{aligned} B(s) &= [1;1.1] \cdot s^2 + [25.8;30] \cdot s + \\ &+ [K_D + 152.8; K_D + 231.4] \end{aligned} \tag{10}$$

After this, a remainder can be derived with the help of (5). Then, adaptation laws (7) can be found:

$$\begin{aligned} K_P(K_D, p_1, p_2, p_3) &= 2 \cdot p_1 - 4 \cdot p_2 + 4 \cdot p_3 + 2 \cdot K_D; \\ K_I(K_D, p_0, p_1, p_2) &= -p_0 + 2 \cdot p_1 - 2 \cdot p_2 + 2 \cdot K_D. \end{aligned} \tag{11}$$

Now, let us choose the value of differential coefficient K_D. To do this, a D-partition method will be applied. Equations of D-partition curves can be derived by substituting (8) and (9) to (10). First, let us find values of K_D, which provide allocation of non-dominant poles inside the desired sector. D-partition curves in vertices of parametric polytope of the system are shown in the figure 3.

IOP Conf. Series: Materials Science and Engineering **707** (2019) 012018 doi:10.1088/1757-899X/707/1/012018

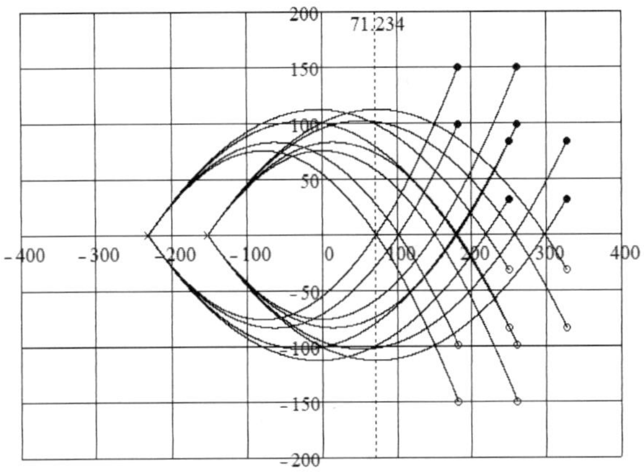

Figure 3. D-partition curves in a plane of K_D, providing allocation of non-dominant poles in the desired sector

From the figure 3 it is clear, that values $K_D = (0;71.234)$ provide desire allocation of non-dominant poles.

Now let us find values of K_D, which provide allocation of non-dominant poles on the left from the line $\text{Re}(x) = -10$ on a complex plane. D-partition curves, in vertices of parametric polytope of the system are shown in the figure 4.

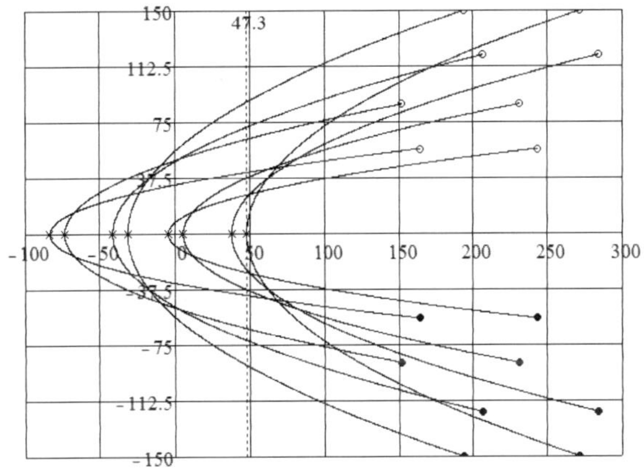

Figure 4. D-partition curves in a plane of K_D, providing allocation of non-dominant poles on the left of desired line

From the figure 4 it is clear, that all values $K_D > 47.3$ provide desired allocation of non-dominant poles. Intersection of intervals from these two D-partitions gives us a set of values $K_D = (47.3;71.234)$, which provides desired allocation of non-dominant poles in a truncated sector. Let us choose $K_D = 50$.

Finally, the controller is synthesized. Expressions for its coefficients can be written as follows:

$$K_P(p_1, p_2, p_3) = 2 \cdot p_1 - 4 \cdot p_2 + 4 \cdot p_3 + 100;$$
$$K_I(p_0, p_1, p_2) = -p_0 + 2 \cdot p_1 - 2 \cdot p_2 + 100;$$
$$K_D = 50.$$

Poles allocation of the synthesized system is shown in the figure 5.

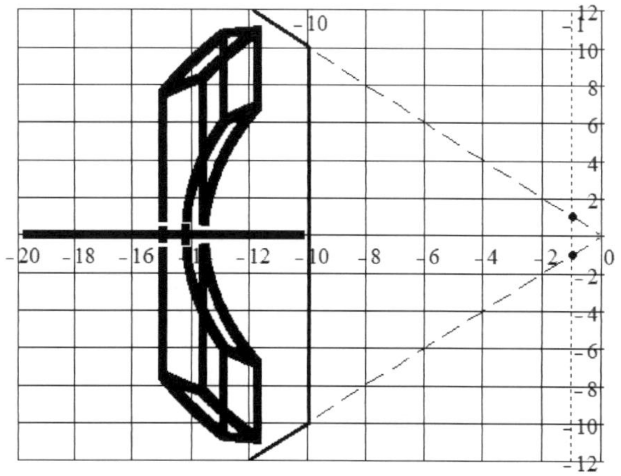

Figure 5. Poles allocation of the synthesized system

From the figure 5 it is clear, that synthesis was successful. Dominant poles are placed in desired points of a complex plane; allocation area of non-dominant poles is placed inside of the desired truncated sector.

In the figure 6 step responses of the synthesized system for different values of interval parameters are shown.

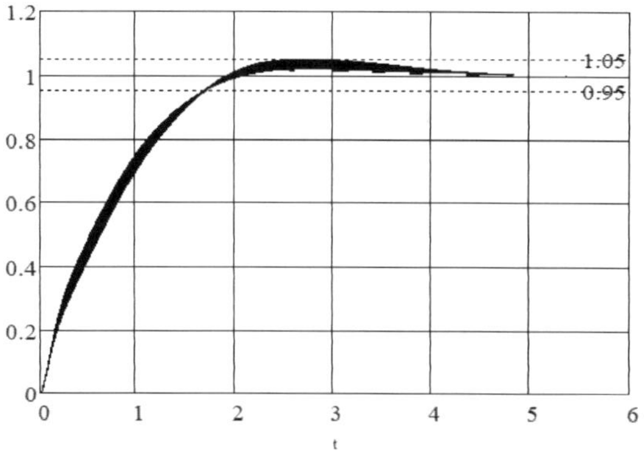

Figure 6. Step responses of the synthesized system

From the figure 6 it is clear, that control quality of the system is sufficient despite interval parametric uncertainty: setting time is equal to 2 seconds; overshoot does not exceed 5%.

6. Simulation modeling

Let us now test the behavior of the synthesized system, when uncertain parameters of the system vary faster, than setpoint signal. To do this, a SIMULINK model of the synthesized system was developed. Diagram of parametric variations during first 90 seconds of modeling is shown in the figure 7; input and output signals of the system are shown in the figure 8.

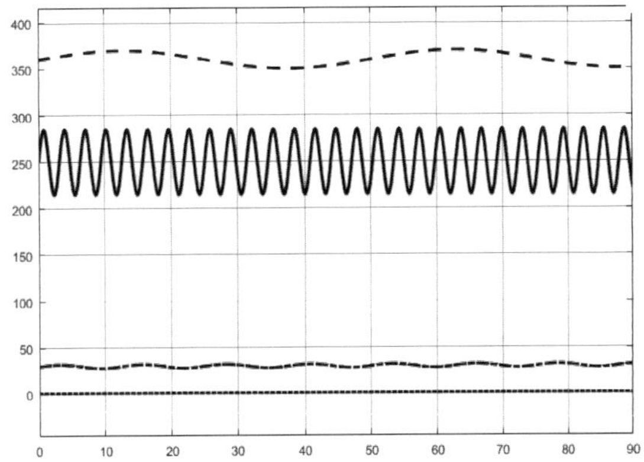

Figure 7. Variation of parameters during simulation modelling of the synthesized

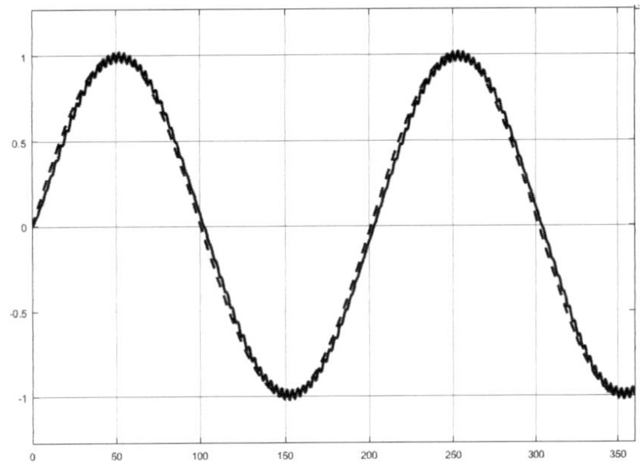

Figure 8. Input and output signals of the synthesized system during simulation modelling

From figures 7 and 8 it is clear, that the system follows the sine input signal accurately. This means, that adaptation laws, synthesized with the help of the method, are effective.

7. Conclusion

The main aim of the research was reached: a new method of synthesis, which allows designing adaptive robust controllers, was developed. Controllers, synthesized with the help of the method, have advantages of interval and adaptive approaches to manipulating systems with uncertain parameters: a procedure of synthesis is simple; a controller structure is simple and strictly defined; a controller is able to damp rapid significant parametric disturbances and provides stable control quality. All these features were proved by simulation modelling in rigid conditions, when parameters of the system vary faster, than input signal does.

8. Acknowledgments

The reported study is supported by the Ministry of Education and Science of Russian Federation (project #2.3649.2017/PCh).

9. References

[1] Ioannou P and Baldi P 2010 *Robust Adaptive Control: The Control Systems Handbook* (Boca Raton: CRC Press).
[2] Ioannou P and Sun J 2012 *Robust Adaptive Control* (United States: Courier Corporation).
[3] Feng G and Lozano R 1999 *Adaptive Control Systems* (Oxford: Newnes).
[4] Wang D, Huang J, Lan W and Li X 2009 Neural network-based robust adaptive control of nonlinear systems with unmodeled dynamics *Math. and Comp. in Sim.* **79(5)** 1745-53.
[5] Lee H, Park J and Chen G 2001 Robust fuzzy control of nonlinear systems with parametric uncertainties *IEEE Trans. on Fuzzy Syst.* **9(2)** 369-79.

[6] Polyak B and Scherbakov P 2002 *Robust stability and control* (Moscow: Science).
[7] Bhattacharyya S, Chapellat H and Keel L 1995 *Robust control: The parametric approach* (United States: Prentice-Hall).
[8] Nesenchuk A 2002 Parametric synthesis of qualitative robust control systems using root locus fields. *Proc. of the 15th Triennial World Congress of The Int. Federation of Automatic Control (Barcelona)* 387-387.
[9] Nesenchuk A and Fedorovich S 2008 Parametric synthesis method of integral systems on the basis of root locus curves of Kharitonov polynomials *Autom. and Rem. Contr.* **69(7)** 1133-41.
[10] Nesenchuk A 2017 A method for synthesis of robust interval polynomials using the extended root locus. *Proc. of the American Control Conf. (Seattle)* 1715-25.
[11] Wang Y, Schinkel M and Hunt K 2002 PID and PID-like controller design by pole assignment within D-stable regions *Asian J. of Contr.* **4(4)** 423-32.
[12] Zhmud V, Dimitrov L and Yadrishnikov O 2014 Calculation of regulators for the problems of mechatronics by means of the numerical optimization method *Proc. of 12th Int. Conf. on Actual Problems of Electronic Instrument Engineering (Novosibirsk)* 739-744.
[13] Zhmud V, Reva I and Dimitrov L 2017 Design of robust systems by means of the numerical optimization with harmonic changing of the model parameters *J. of Phys.: Conf. Ser.* **803(1)**.
[14] Tagami T and Ikeda K 2003 Design of robust pole assignment based on Pareto-optimal solutions *Asian J. of Contr.* **5(2)** 195-205.
[15] Gayvoronskiy S and Ezangina T 2014 The synthesis of the robust stabilization system of cable tension for the test bench of weightlessness simulation *Adv. Mat. Res.* **1016** 394-99.
[16] Gayvoronskiy S, Ezangina T and Khozhaev I 2015 The interval-parametric synthesis of a linear controller of speed control system of a descent submersible vehicle *IOP Conf. Ser.: Mat. Sc. and Eng.* **93**.
[17] Gayvoronskiy S, Ezangina T and Khozhaev I 2017 Method of interval system poles allocation based on a domination principle *Int. Conf. on Mechanical, System and Control Engineering* (St. Petersburg) 245-249.

IOP Conf. Series: Materials Science and Engineering **707** (2019) 012019 doi:10.1088/1757-899X/707/1/012019

Characterization of damages in materials by computer-aided tap testing

Imran Oral[1]

[1]Department of Physics Education, Necmettin Erbakan University, 42090 Konya, Turkey.

E-mail: oralimran@erbakan.edu.tr

Abstract. The tap test is an easy and effective way of characterization of flaws and damages in different materials (especially honeycomb sandwich composites). When the tapping is carried out systematically to cover an area, the measured impact duration can be used for generating a "scan image" of the tapped area. Such an instrumented tap test and imaging system can be used for the detection of flaws and damages, for the evaluation of repaired areas and for the verification of normal reinforced substructures. In this study, the images of intact and damaged aluminum honeycomb sandwich composite, pine board and poplar board samples were obtained by using Computer Aided Tap Tester (CATT) system in the electric and magnetic field and also without any field. As a result of the tests made, it has been figured out that the tests made in the electric field and the magnetic field increased the quality of the images for damages. Also, it has been found out that the CATT system is quite successful in detecting damage of honeycomb sandwich composite materials.

1. Introduction

Composite materials are used in many applications ranging from aerospace vehicles to consumer products in our daily life. The composite structures can be damaged from many causes. For instance, these composite materials can be damaged during the production process and during the in-service life of them [1]. For many composite structures, the most significant damages occur by mechanical impacts. Impact damage can be created by dropped tools, hailstone and bird strikes etc. Composite materials have a variety of damage types such as delamination, micro-cracking, and porosity. Therefore, important materials such as composite structures used in aircraft's bodies need to be able to quickly pass through non-destructive inspection process. Nowadays, the development of health monitoring solutions has gained considerable attention [2-4].

The hearing-based tap test is one of the most practiced inspection technique on composites. This method only requires a coin and a good sense of hearing. When the surface of a material is damaged with the coin, it creates a sound which is related with the local stiffness. If a tap occurs over a damaged zone, one can hear a sound have a "dead" tone. While a tap occurs over an undamaged zone, one can hear a sound have a higher tone[5]. However, the hearing-based tap test depends on the human ear. There are three developments of instrumented tap test which eliminated the human factor of reliance on operator hearing. One of this system, which eliminates the dependence of human ear is developed by Iowa State University and is called the Computer Aided Tap Tester (CATT).

The computer aided tap test is a method, which makes use of an accelerometer and electric circuit to capture the contact time of the tap. The amplitude of the acceleration depends on the force used along the tapping. However, the contact time is independent of this force. Based on a simple grounded spring harmonic oscillation model, the contact time τ is related to the local stiffness (spring constant) k by the simple equation-1[5]:

$$k = m\left(\frac{\pi}{\tau}\right)^2 \qquad (1)$$

Where k, m and τ are stiffness, mass of the tapper and contact time respectively. The stiffness at different regions tested part can then be plotted as a stiffness image of the composite structure. Therefore, the CATT system can quickly figure out the damaged regions in the materials. In the related literature there is not any study, which has figured out the effect of electric and magnetic fields on images of damage obtained by CATT system. Therefore, this research was done for investigating the effect of electric and magnetic fields on the characterization of damage in materials by Computer Aided Tap Tester (CATT).

2. Experimental

2.1. Materials and Equipment

In this research a Computer aided tap tester (CATTV5_3-ASI, USA) system, an aluminum honeycomb sandwich panel-15cm x 15 cm x 1 cm - (Altıgen, Turkey), a poplar board (13,5 cm x 13,5 cm x 2,4 cm) and a pine board (15 cm x 15 cm x 2,3cm) were used. Also, in the experiments strip and neodymium magnets, 1,5 V DC power supply, multimeter, rheostat, aluminum foil, transparent 0,5 cm partitioned acetate, tape, ruler with mm section and conductive cable were used.
Sandwich materials used in the research are produced by Altıgen Space Aviation, Ship Construction Industry Trade Ltd. Co. The materials used in the research and the experimental setups are given in Figure1.

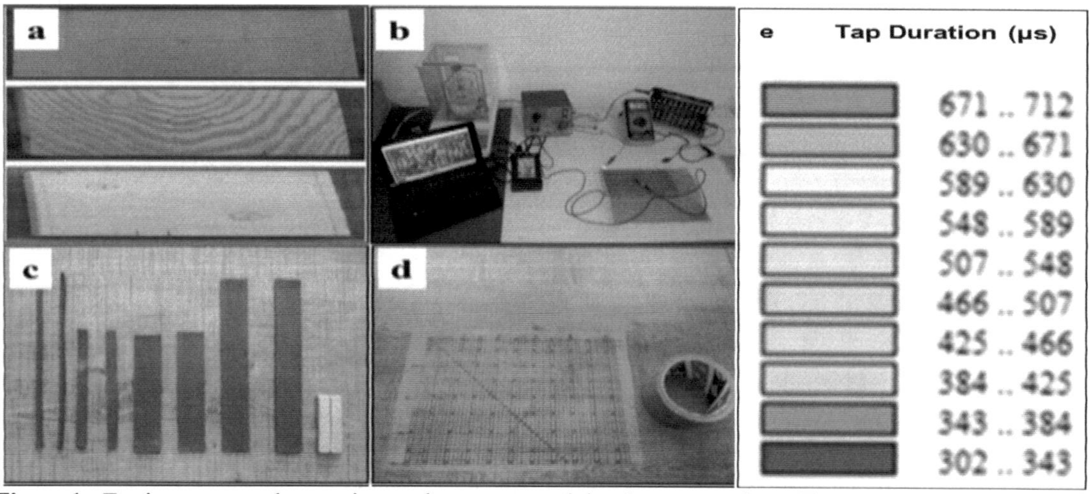

Figure1. Equipments and experimental set up used in the research: a-The materials used in the research, b-CATT system and electric field circuit, c- Strip and neodymium magnets, d- Tape and 0.5 cm partitioned acetate, e- Tap duration time scale.

First of all, the contour images of intact samples obtained with the CATT system and then all materials samples were damaged. The damages of the samples used in this study were formed by drill machine and a hammer.

2.2. Materials Characterization with Computer Aided Tap Tester (CATT)

In this research, the contour images of intact aluminum honeycomb sandwich, poplar board and pine board were obtained by using Computer Aided Tap Tester (CATT) system in the electrical and magnetic fields and also without any field. The contour images were obtained using tap duration times (contact time) and stiffness values of the samples.

After opening the ASI CATTV5_3 program installed on the Acer one 10 model computer, a scanning window equivalent to the size of the sample examined is opened. When this window is opened, the unit of inch, cm or mm and the step of progression can be selected at the desired accuracy (eg 0.5; 1; 2 or 2.5 steps if cm is selected). For each tapping operation, the contact time of accelerometer appears in different colours in microseconds (μs) on the screen of the computer. The obtained contact time values can be transformed into the local stiffness contour image with CATTV5_3 program in accordance with the equation of 1, and CATTV5_3 can also give 3D images of these contact times and stiffness values using the data.

3. Findings

The normal contour images obtained from the contact time and stiffness values of all the materials used in the research were obtained by tapping tests without any field and in presence of electric field and magnetic field. These images are given in Figures 2-7.

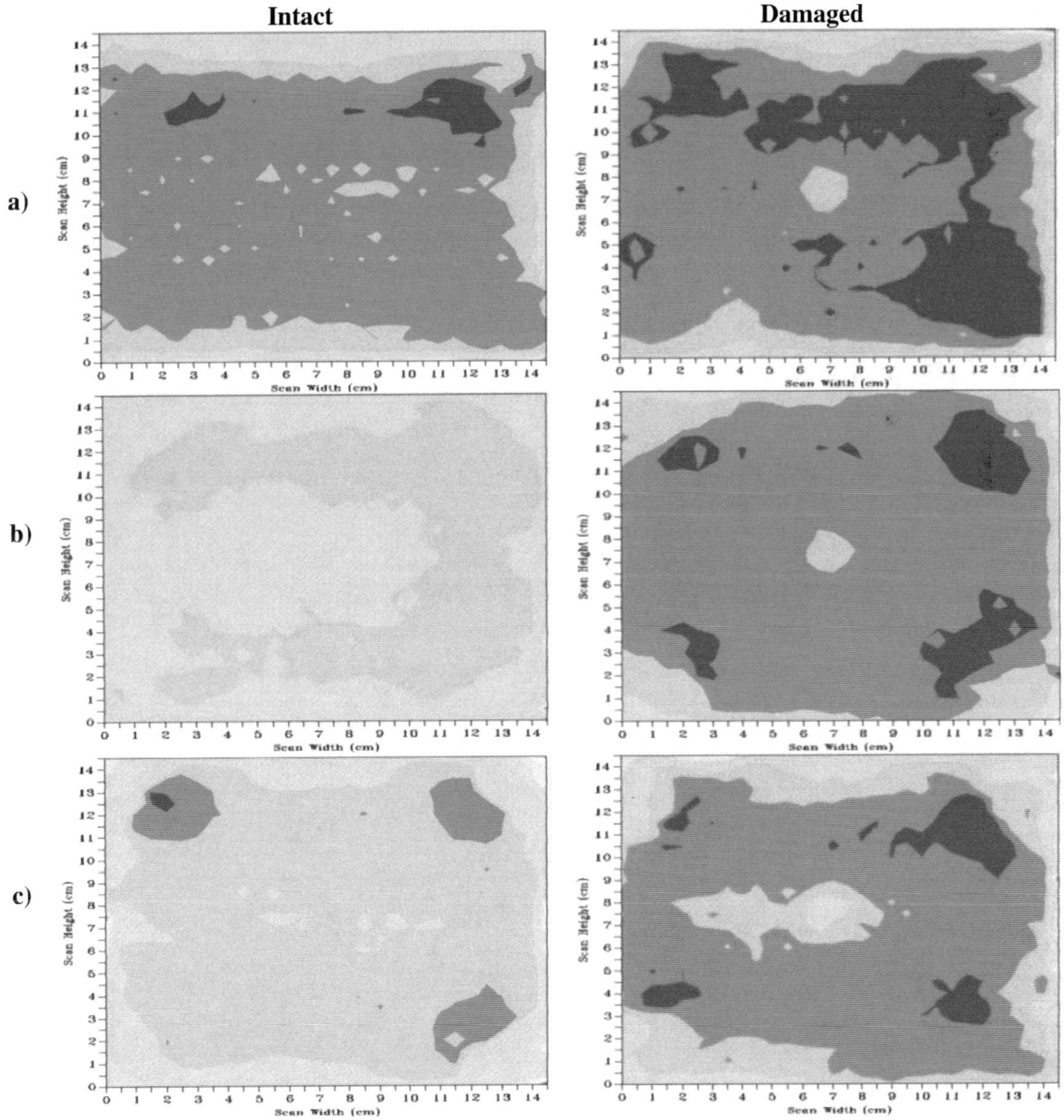

Figure2. The images of intact and damaged aluminum honeycomb sandwich panel obtained by using the contact time: a) Without any field, b) With electric field, c) With magnetic field.

a)

b)

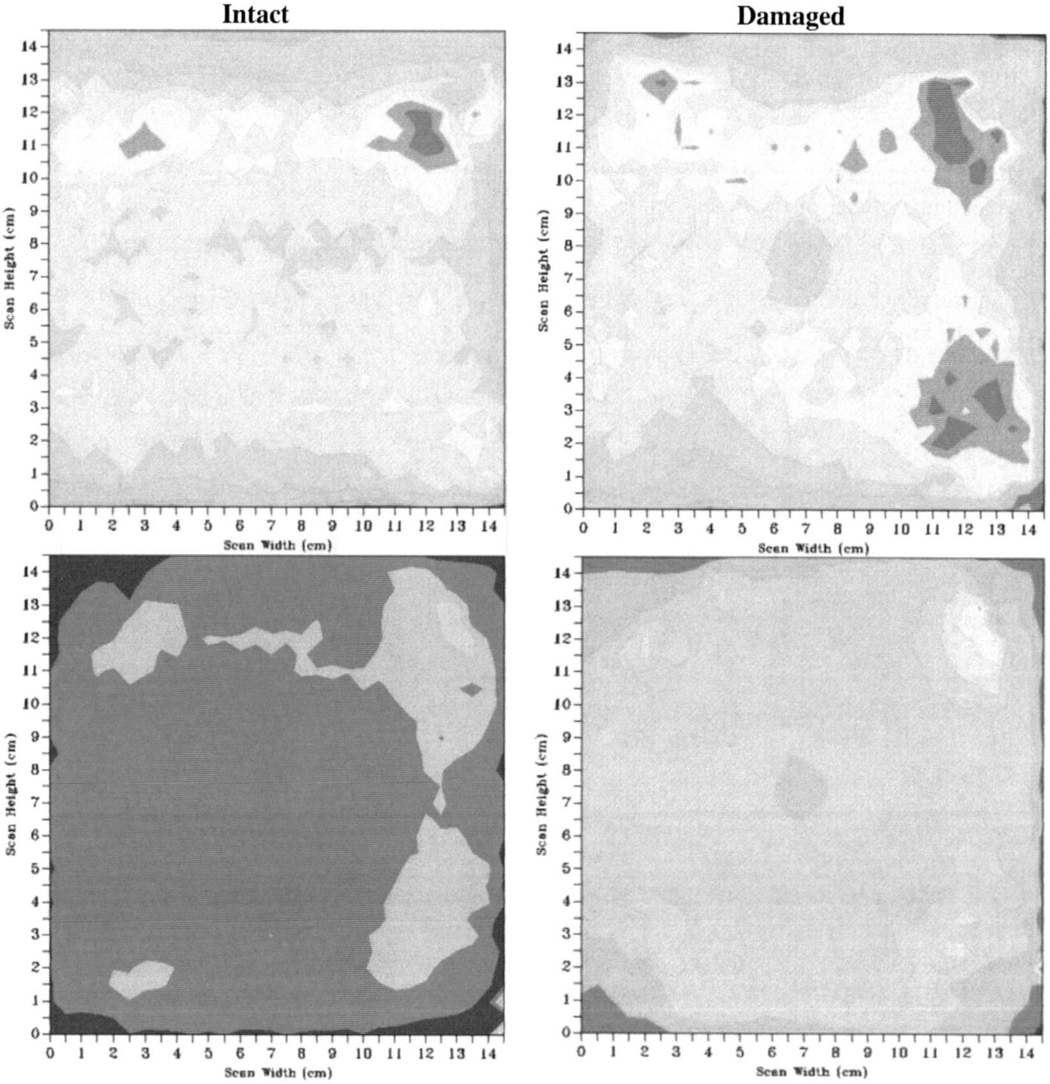

IOP Conf. Series: Materials Science and Engineering **707** (2019) 012019 doi:10.1088/1757-899X/707/1/012019

c)

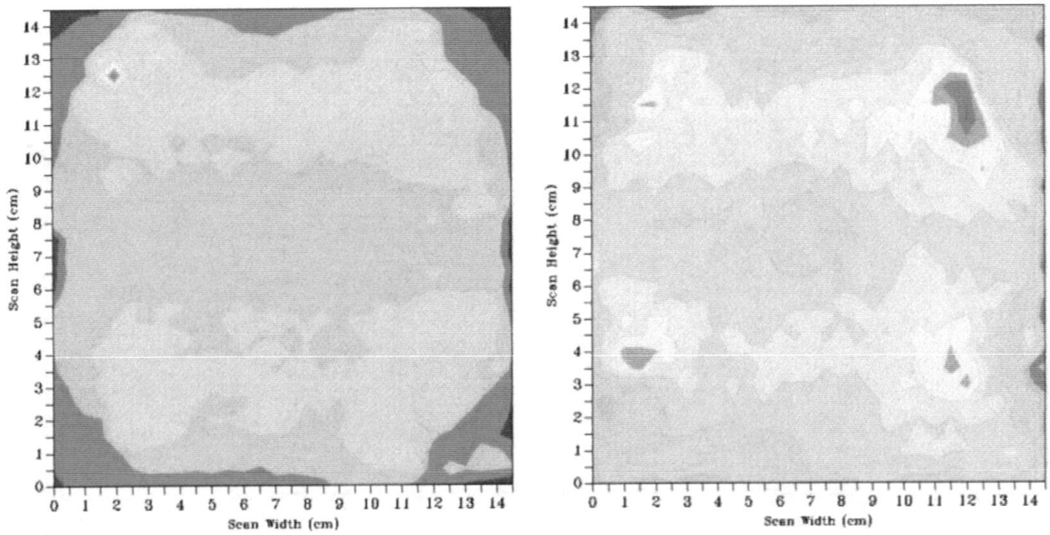

Figure3. The images of intact and damaged aluminum honeycomb sandwich panel obtained by using the stiffness values: a) Without any field, b) With electric field, c) With magnetic field.

a)

b)

c)

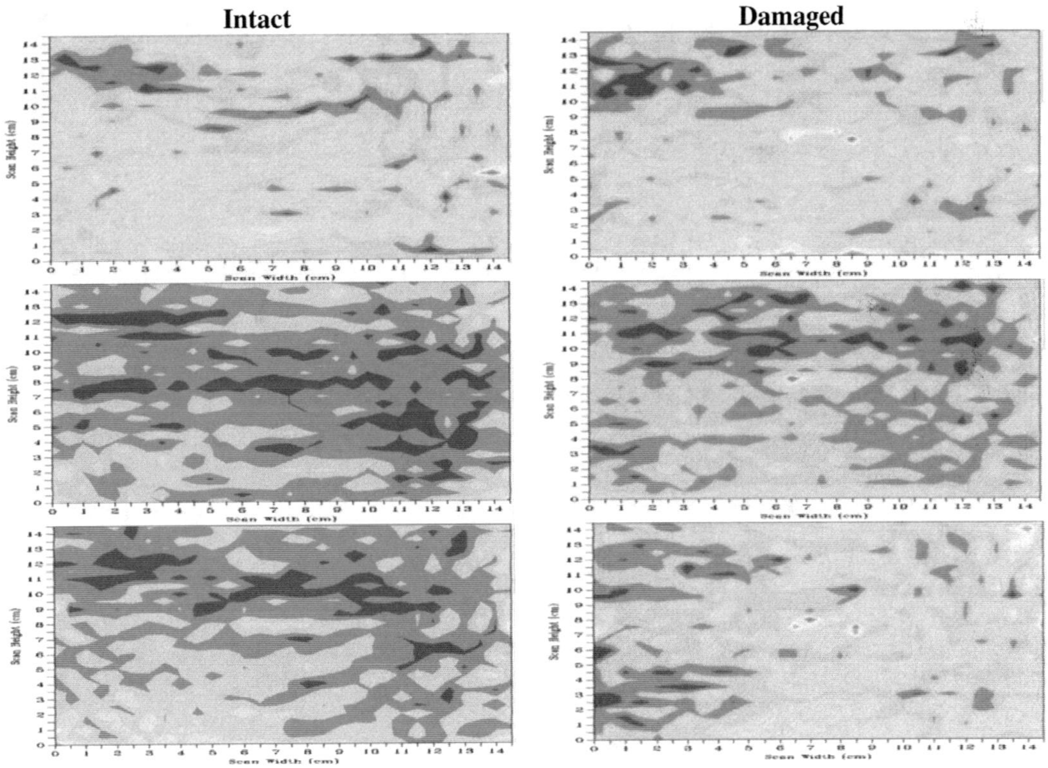

Figure4. Images of intact and damaged pine board obtained by using the contact time: a) Without any field, b) With electric field, c) With magnetic field.

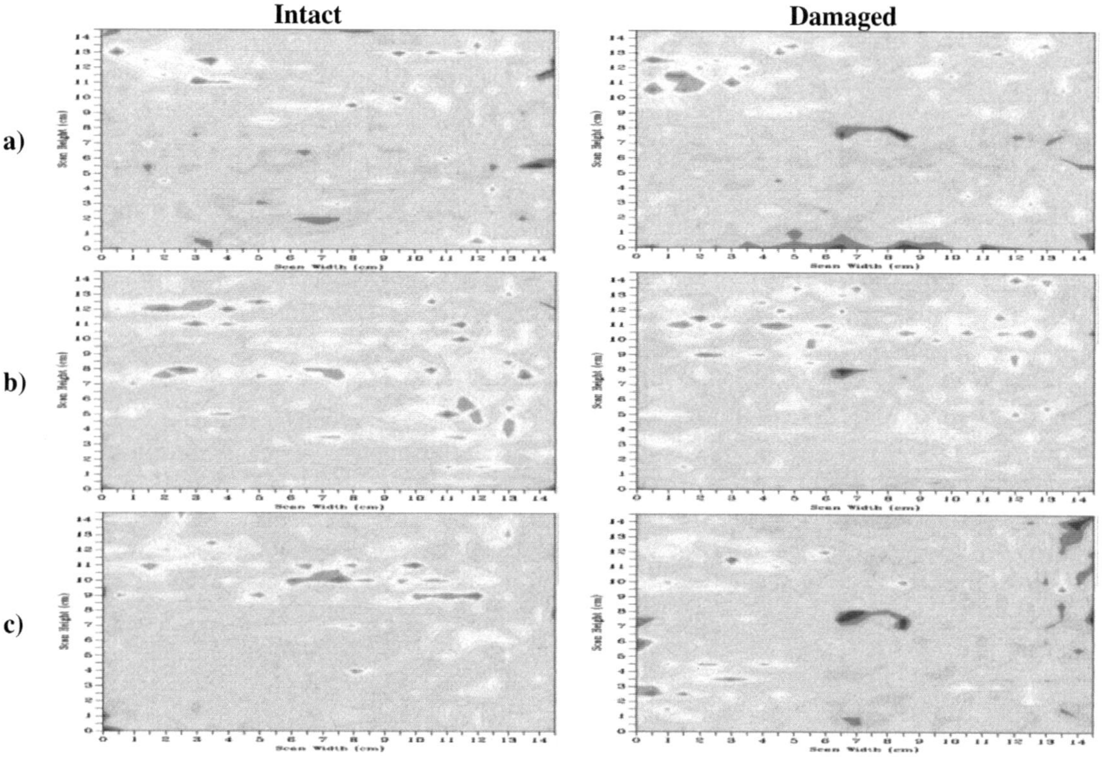

Figure5. Images of intact and damaged pine board obtained by using the stiffness values: a) Without any field, b) With electric field, c) With magnetic field.

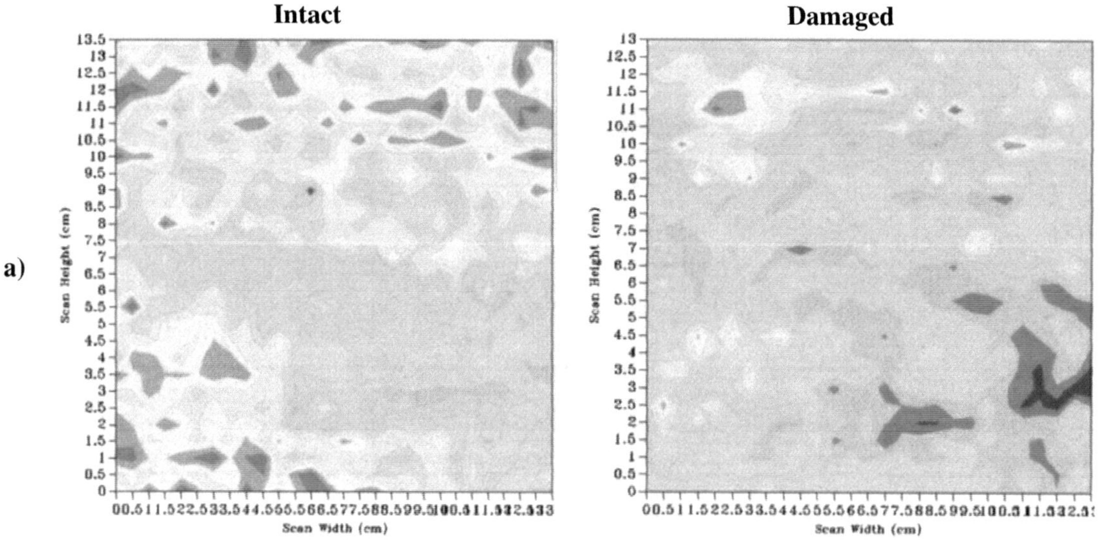

b)

c)

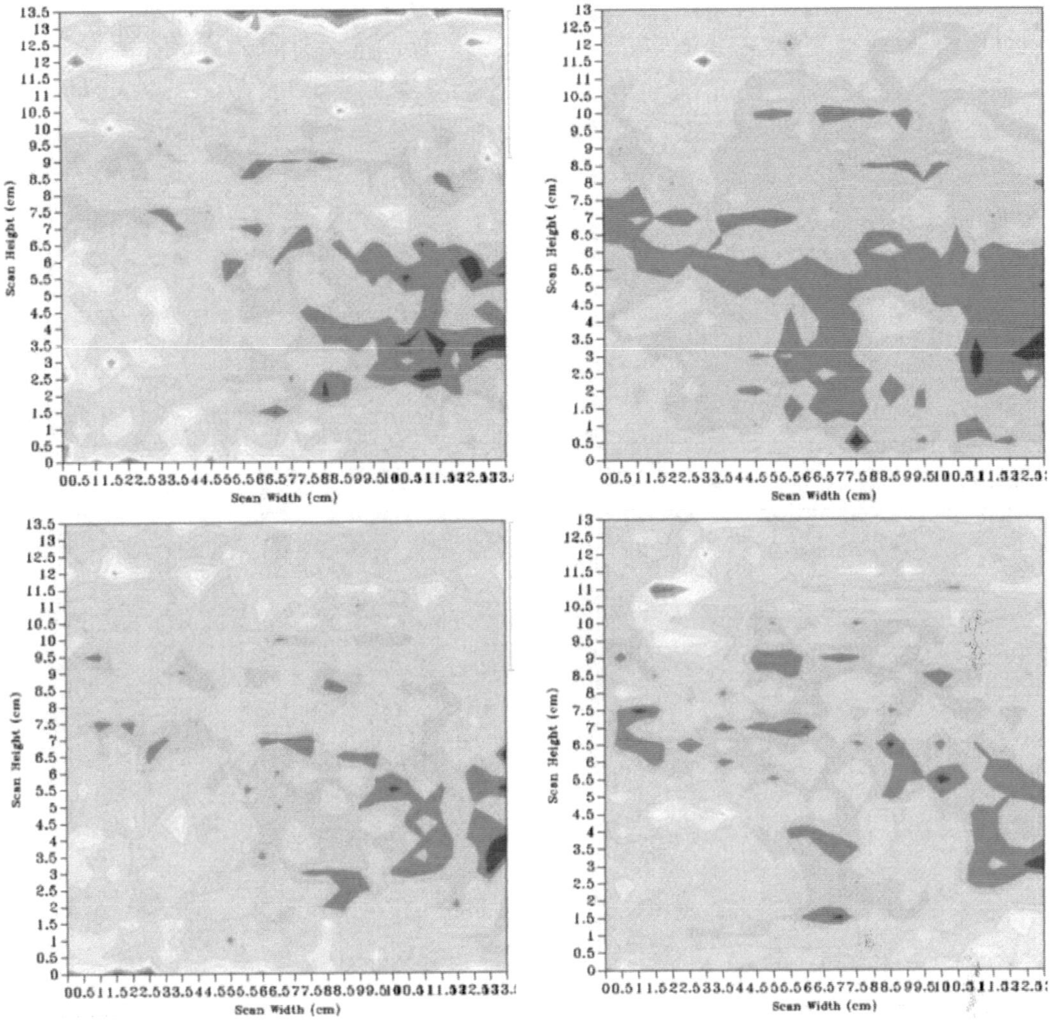

Figure6. Images of intact and damaged poplar board obtained by using the contact time: a) Without any field, b) With electric field, c) With magnetic field.

Intact **Damaged**

a)

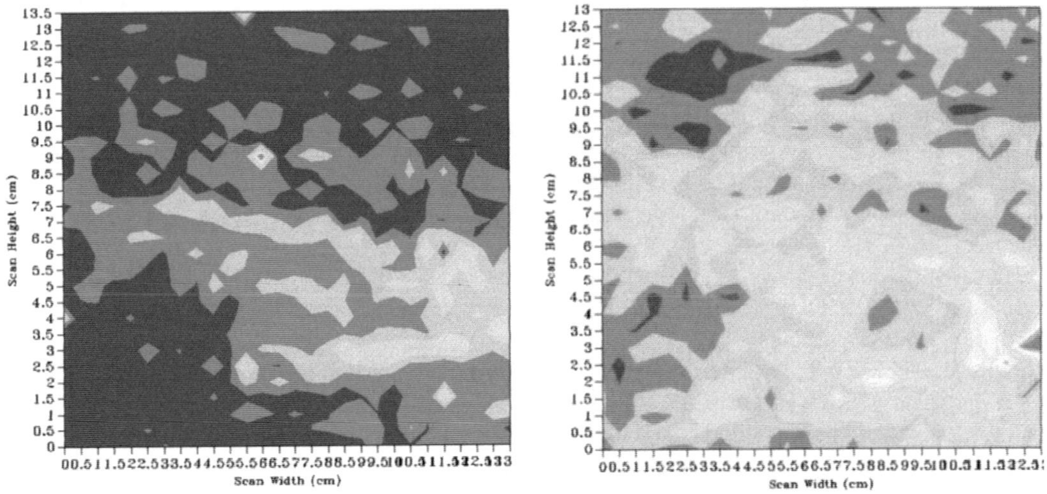

b)

c)

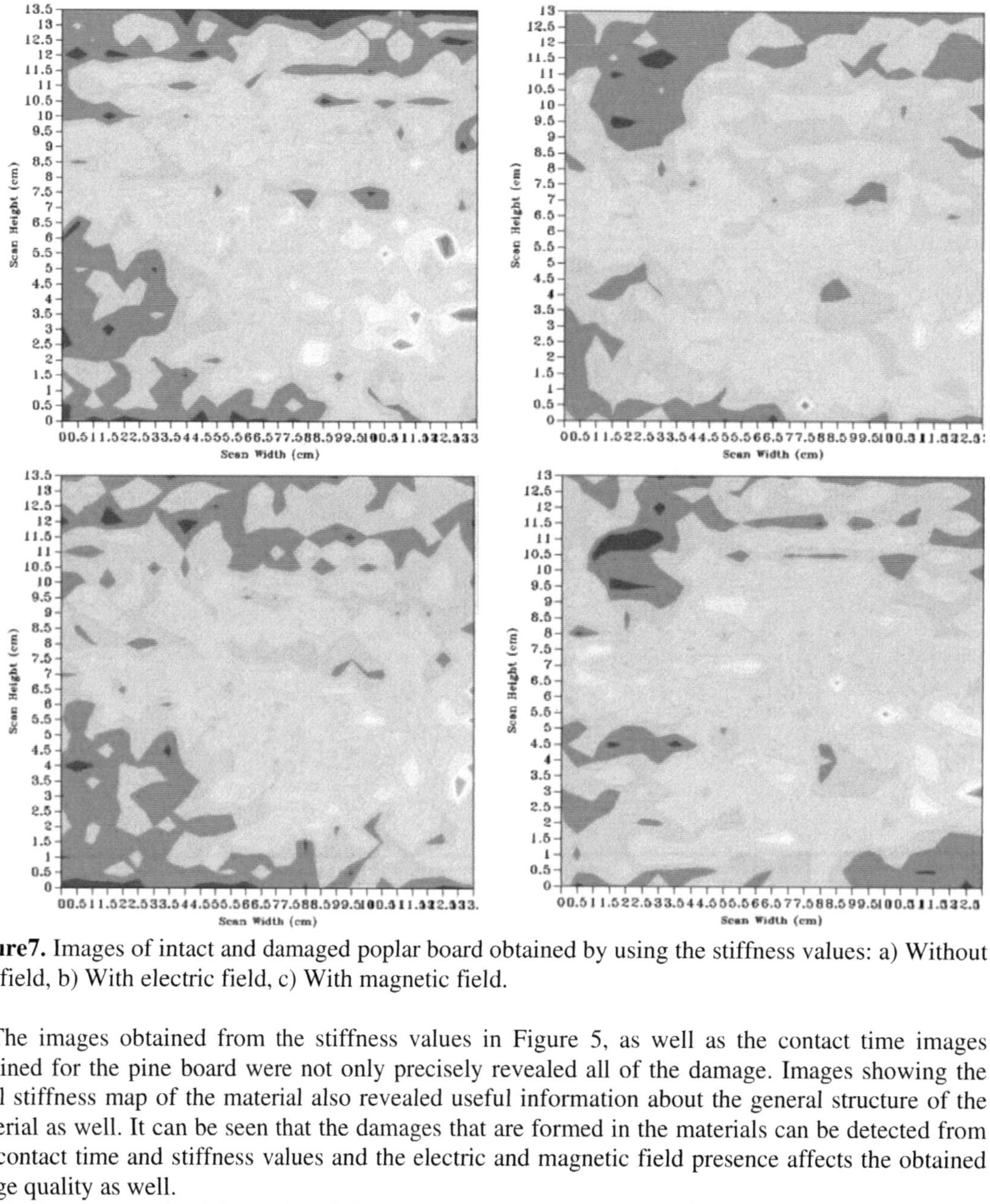

Figure7. Images of intact and damaged poplar board obtained by using the stiffness values: a) Without any field, b) With electric field, c) With magnetic field.

The images obtained from the stiffness values in Figure 5, as well as the contact time images obtained for the pine board were not only precisely revealed all of the damage. Images showing the local stiffness map of the material also revealed useful information about the general structure of the material as well. It can be seen that the damages that are formed in the materials can be detected from the contact time and stiffness values and the electric and magnetic field presence affects the obtained image quality as well.

The images obtained from the stiffness values in Figure 7, as well as the contact time images obtained for the poplar board were precisely revealed all of the damage. Especially, the images showing the local stiffness map of the material revealed the damage very well. The images give very important information related the damages on the surface and subsurface of the material and the structure of the material.

4. Results and Discussions

The product range in the manufacturing sector; quality, technical safety, durability and suitability for the purpose of use, as well as the unconditional assurance of the product is formed with the developing technology. As the aim is to provide continuous and high quality in production, the importance of non-destructive inspection methods in quality control applications is increasing. As described before, there are many non-destructive test methods.

The physics of the coin-tap method have been very well investigated and explained by [6]. The Wichitec RD3 made of Wichitech Industries, the Computer-Aided Tap Testing (CATT) made of Advanced Structural Imaging Inc. [7] and the Woodpecker made of Mitsui Technologies are examples of the automated tap testing systems that eliminate the reliance the human hear. It was figured out that the load versus displacement curve of an un-damaged composite was linear and there was not any hysteresis. Also, the average slope of the load-displacement stiffness of an undamaged composite honeycomb panel was higher than the damaged composite sample. It was also found out that the area enclosed by the hysteresis loop gave a good correlation with the impact energy which caused the damage on the composite sample [8]. On the other hand, Gryzagoridis and Findeis [9], had used Digital Shearography and Tap Testing Methods to figure out the damage in composite materials of both the monolithic and sandwich types. They used the Digital Shearography to benchmark the results obtained with the Tap Testing Method. They had performed the tap testing using the Mitsui Woodpecker WP-632AM which is a hand held automated impact instrument that displayed the defects' positional data on a PC in real time. They had figured out that the results obtained with Digital Shearography on the same test sample had proven the results obtained by the Tap Testing Method.

In this research, the Computer Aided Tap Tester (CATT) is used to figure out the damage by without using any field and using electric and magnetic fields. As it can be seen from the related literature, there are not any researches that have investigated the effect of the electric and magnetic fields to reveal the damage of materials. Therefore, this research is a unique research. Because that this is the first time that the effect of the electric and magnetic fields on the damage characterization and on tap test images is studied.

Within the scope of the research, the damage generated in various materials was tried to be determined with contour images obtained from the stiffness values and contact times of tapper on the surface of the materials to investigate the electric field and magnetic field effect by CATT system. Thus, the effect of the electric and magnetic fields on the damage have been investigated with a new method, which has never been tried before in the literature.

It can be seen from the Figure 2 and Figure 3 that both the contact time and stiffness values can be used to determine the damage of the material and at the same time the electric and magnetic field presence affects the obtained image quality. A difference in images in the presence of electric and magnetic fields was an expected result. Because the aluminum honeycomb sandwich panel's core and face sheet is made of aluminum. The main reason underlying the expectation related this difference is that aluminum is a material that has both conductive and paramagnetic properties. For paramagnetic materials, the direction of magnetization is the same as the direction of the external magnetic field. As a result of this, paramagnetic materials are attracted by magnets [10]. Materials such as aluminum can get a different structure if they are transient due to the effects of electric dipoles and magnetic dipoles in the magnetic field. For this reason, it can be said that the images obtained from the tapping test made on the damaged aluminum and intact aluminum materials in the presence of electricity and magnetic field are obtained with better quality.

The contour images obtained by using the contact time and stiffness values of the pine board are shown in Figure 4 and Figure 5. From the data of these figures, it can be seen that the damage in pine board can be clearly revealed from the contact time and stiffness values as seen in aluminum honeycomb sandwich panel materials. Also, at the same time the presence of electric and magnetic fields affect the image quality positively as well. The reason for this is that the stiffness of the pine board is lower than the stiffness of the tapper.

According to the Figure 6 and Figure 7, which obtained for intact and damaged poplar board, it can be concluded that the damage can be sensitively revealed by the CATT system in the presence of the electric and magnetic fields as well as aluminum and pine board samples. According to this result, it can be said that the CATT system is a very successful, practical, easy and inexpensive method for revealing the structures of wooden structures and their damages in particular.

5. Conclusions

Based on the experimental results of this study, which is carried out by the CATT system, the following conclusions can be drawn:

1-The damage in honeycomb sandwich panel materials can be figured out using CATT system without using any field. However, when tapping test was carried out in the presence of electric and magnetic fields, the damages in these materials can be revealed more sensitively.

2- The effect of electric and magnetic fields has been tested for the first time in tap test by using the CATT system.

3- It was found out that the CATT system is very suitable for revealing the damage in honeycomb sandwich composite panels.

In this study, the presence of electrical and magnetic fields was tested to see whether there was any effect on the images obtained by the tapping test in order to determine the damages in various materials. This research is so important by investigating the effect of electric and magnetic fields in damage analysis by tapping test for the first time. The results of the research showed that electrical and magnetic fields have different effects in damage analysis with the tapping test. For this reason similar studies can be repeated with different electric and magnetic fields levels. Damage caused by impact tests with different energies can be analysed by the CATT system in the presence of different magnitudes of electrical and magnetic fields too.

6. Acknowledgments

This study was supported by the Scientific Research Projects (BAP) Coordinating Office of Necmettin Erbakan University by the project number of 161210018. The author is grateful for the support provided by Necmettin Erbakan University, Turkey.

7. References

[1] Bowkett M, Thanapalan K and Williams J 2016 Review and analysis of failure detection methods of composites materials systems. In: 2016 22nd International Conference on Automation and Computing, ICAC 2016: Tackling the New Challenges in Automation and Computing, pp 138-43.

[2] Bowkett M and Thanapalan K 2017 Comparative analysis of failure detection methods of composites materials' systems Systems Science and Control Engineering 5 168-77.

[3] Cheng L and Tian G Y 2012 Comparison of nondestructive testing methods on detection of delaminations in composites Journal of Sensors 2012.

[4] Luo Y, Wang Z, Wei G, Alsaadi F E and Hayat T 2016 State estimation for a class of artificial neural networks with stochastically corrupted measurements under Round-Robin protocol Neural Networks 77 70-9.

[5] Hsu D K, Barnard D J, Peters J J and Dayal V 2000 Physical basis of tap test as a quantitative imaging tool for composite structures on aircraft Aip Conf Proc 509 1857-64.

[6] Cawley P, Adams, R.D. 1988 The Mechanics of the Coin-Tap Method of Non-Destructive testing Journal of Sound and Vibration 122 299-316.

[7] Imaging A S 2018 Using the CATT.

[8] Foreman C D 2008 Nondestructive detection and characterization of damage in honeycomb composite structures. In: Aerospace Engineering, (Ames, Iowa: Iowa State University).

[9] Gryzagoridis J, Findeis, D. 2013 Tap Testing of Composites Benchmarked with Digital Shearography. Retrieved June 22, 2018, from http://www.ndt.uct.ac.za/sites/default/files/image_tool/images/111/BINDT2013A.pdf

[10] Keller F J, Gettys, W. E., Skove, M. J. 1993 Physics II (New York: McGraw-Hill).

AUTHOR INDEX